The Way of Mountains and Rivers

The Way of Mountains and Rivers

Teachings on Zen and the Environment

With Commentary on Zen Master Dogen's Mountains and Rivers Sutra

John Daido Loori

Published by
Dharma Communications

For information, write Dharma Communications at
P.O. Box 156, 831 Plank Road, Mt. Tremper, NY 12457

9 8 7 6 5 4 3 2 1

First Edition
Printed in China
♻Matt Kote Green 100% Recycled Artpaper

Library of Congress Number: 2009920392

ISBN 978-1-88279-521-5

This book is dedicated to our fragile planet.

May we discover the skills necessary to live in peace and harmony with all beings.

Although we may have eyes to see the mountains as the appearance of grass and trees, earth and stone, fences and walls, we should understand that this is not yet its total manifestation: it is not the complete actualization of the mountains.

—Zen Master Dogen

Contents

Acknowledgements vii

Introduction ix

About the Photographs xii

Master Dogen's *Mountains and Rivers Sutra* 2

1. These mountains and rivers of the present . . . 10
2. It is because the blue mountains are walking that they are constant. . . . 16
3. We should realize that the blue mountains must be understood . . . 22
4. The blue mountains commit themselves to the practice of walking . . . 28
5. Although we may have eyes to see the mountains . . . 34
6. "The stone woman gives birth to a child in the night." . . . 40
7. At the present time in the land of Song . . . 46
8. We should understand that the teaching of "the East Mountain" . . . 52
9. Water is neither strong nor weak, neither wet nor dry . . . 58
10. In general, there are different ways of seeing mountains and rivers . . . 64
11. Therefore, what different types of beings see is different. . . . 70
12. Thus, water is not earth, water, fire, wind, space, or consciousness . . . 76
13. The Wenzi says, "The Tao of water is to ascend to the sky" . . . 82
14. Furthermore, we should not think that when water has become . . . 88
15. However, when dragons and fish see water as a palace . . . 94
16. From the timeless beginning to the present, the mountains . . . 100
17. An ancient sage has said, "If you wish to avoid the karma of hell" . . . 106
18. Throughout history, we find many examples of emperors . . . 112
19. Since ancient times, wise ones and sages have also lived by the water. . . . 118
20. Therefore, water is the palace of the true dragon. . . . 124

Index of Photographs 130

Glossary 131

About John Daido Loori 133

About Zen Mountain Monastery 133

Acknowledgments

First and foremost, I must express my gratitude to Eihei Dogen, whose teachings are a constant source of inspiration and a deep well of wisdom.

I am grateful to Vanessa Zuisei Goddard for her skillful editing and to Konrad Ryushin Marchaj Sensei for his counsel. Thank you also to Joanne Dearcopp and Chase Twichell for their careful reading of the manuscript and their helpful advice, and to William Tendo Leckie for his graphics and layout work.

I also wish to acknowledge Prof. Carl Bielefeldt, David Sanshin Noble and Andrew Hobai Pekarik, whose previous translations of the *Mountains and Rivers Sutra* served as the basis for the contemporary English version presented here.

Finally, I wish to thank the many participants in the intensive "Mountains and Rivers" workshops offered at Zen Mountain Monastery in 1996 and 2006. Their questions and comments helped to shape the current translation and commentary, making Dogen's teachings more accessible to the western reader.

Introduction

Dogen was doing zazen in a dark and quiet zendo. In the stillness and silence of the early morning, Rujing bellowed at one of the monastics, "When you study under a master, you must drop body and mind. What's the use of single-minded intense sleeping?!" Sitting right beside this monastic, Dogen suddenly attained great enlightenment. Immediately, he went up to the abbot's room and burned incense. Rujing said, "Why are you burning incense?" Dogen said, "Body and mind have dropped off." Rujing said, "Body and mind dropped off. The dropped-off body and mind." Dogen said, "This may only be a temporary ability. Please don't approve me arbitrarily." Rujing said, "I am not." Dogen said, "What is that which isn't given arbitrary approval?" Rujing said, "Body and mind dropped off." Dogen bowed. Rujing said, "The dropping off is dropped."

Body and mind fallen away is a realm in which there are no doctrines or marvels, no certainties or mysteries. It is just that when you see it, there is not a single thing. Having reached this place, Dogen was able to express it. Rujing approved and Dogen bowed. Having passed through the forest of brambles, Dogen then passed beyond the other side. This is why Rujing said, "The dropping off is dropped."

Thirteenth-century Japanese Zen Master Eihei Dogen is one of the most remarkable religious figures and teachers in the history of Zen. Although relatively unknown during his lifetime, he is now highly regarded as a great philosopher, mystic and poet, both in Japan and in the West.

Dogen became ordained at the age of thirteen in the Tendai monastic complex on Mount Hiei outside Kyoto, then later entered the Zen Monastery Kenninji. There, the priest Eisai, having just returned from China, was introducing the teaching and practice methods of Chinese Zen that he had learned from various teachers of the Linji or Rinzai school of Zen Buddhism and Dogen became steeped in these teachings. After Eisai's death, Dogen continued his studies under Eisai's successor, Myozen, and nine years later traveled with him to China, where he met Tiantong Rujing. Impressed by Rujing's insight and discipline in practice, Dogen remained with him and practiced rigorously until, as the story above describes, he achieved enlightenment. Shortly thereafter, Rujing transmitted the dharma to Dogen and passed away the following year. Dogen then returned to Japan, where he again took up residence at Kenninji, this time as a certified teacher in the Caodong or Soto lineage of Zen. Dogen lived in the Kyoto area for some fifteen years, then established Eiheiji temple in the remote mountains of Echizen.

The teachings of Dogen have come down to us from those years spent at Kenninji and Eiheiji, most notably in his masterwork *Shobogenzo: Treasury of the True Dharma Eye,* a collection of ninety-five extensive discourses on various aspects of the dharma, among them *Sansuikyo: The Mountains and Rivers Sutra*. When Dogen began working on this sutra in the year 1240, he was at the pinnacle of his literary and teaching powers. In fact, many consider this fascicle to be the most eloquent of his writings.

It is impossible to study the *Mountains and Rivers Sutra* and not be moved by the poetry and creativity of Dogen's words. His way with language is so unusual that among scholars it has earned the appellation "Dogenese." He communicates not only with ordinary language, but also with what he calls "intimate words," *mitsugo. Mitsu* means "secret" or "mystical." It is something not apparent to the senses or intellect, which means it requires direct, immediate, and complete perception. *Go* means "words," so mitsugo is "secret talk," or "secret words." It is expression that is communicated directly, but without sound. That is, they are "words without words." Because of their nature, mitsugo are said to be "turning words" that are grasped in a moment of insight, rather than through linear, sequential thought.

Like mitsugo, koans or "public cases" are exchanges that open us up to the possibility of looking at a question directly and intimately. Because they cannot be addressed intellectually, koans cut through the layers of conditioning we have accumulated since birth. In this way, they help us to directly grasp our inherent nature so that we can then live our lives out of that which has been realized.

In this sutra Dogen uses metaphors, mitsugo, traditional koans, and the Five Ranks of Master Dongshan to take up, once again, a theme he had begun exploring in a previous fascicle called *Keiseisanshiki: The Sound of the Stream and the Form of the Mountain.* In *Sound of the Stream,* Dogen first equates the mountains and rivers with the body of the Buddha, while in the *Mountains and Rivers Sutra* he focuses specifically on the meaning of mountains and rivers as the teachings and speech of the Buddha. He points out in his opening lines that the environment that surrounds us right here and now is the expression of the ancient buddhas—is, in fact, a sutra, the word of the Buddha. The mountains and rivers themselves are a teaching that reflects the truth of the buddhadharma.

Another central theme of this fascicle is the Buddhist teaching of nonduality: the realization of no separation between self and other, between self and the rest of the universe. Dogen's teachings on nonduality are based on the *Flower Garland Sutra,* a Mahayana text that describes reality in terms of the fourfold *dharmadhatu* or dharma realms. In the Caodong school, this teaching is presented as the Five Ranks of Master Dongshan. Used within the context of Zen training and the teacher-student relationship, the Five Ranks form a framework for the interplay between the absolute and relative realms.

The first rank is "the relative within the absolute" or emptiness—no eye, ear, nose, tongue, body, or mind. The second rank is the realization of emptiness, and is referred to as "the absolute within the relative." This is where the enlightenment experience, or *kensho,* occurs. Yet absolute and relative are still dualistic. The third rank is "coming from within the absolute." No longer in the abstract, the whole universe becomes each of our lives, and inevitably, compassion arises. Dongshan's fourth rank is "arriving at mutual integration," the emergence from both absolute and relative. At this stage, both realms are integrated, but they are still two realities. In the fifth rank, "unity attained," there is no more duality—neither absolute nor relative, up nor down, profane nor holy, good nor bad, male nor female.

In one example of his use of the Five Ranks, Dogen writes:

> ...Since ancient times wise ones and sages have also lived by the water. When they live by the water they catch fish or they catch humans or they catch the Way. These are traditional water styles. Further, they must be catching the self, catching the hook, being caught by the hook, and being caught by the Way.

Then he introduces a koan to illustrate his point:

> In ancient times, when Chuanzi suddenly left Baishan and went to live on the river, he got the sage Jiashan of the Flowering River. Isn't this catching fish, catching humans, catching water? Isn't this catching himself? The fact that Jiashan could see Chuanzi is because he is Chuanzi. Chuanzi teaching Jiashan is Chuanzi meeting himself.

The phrase "When Jiashan sees Chuanzi, he is Chuanzi," refers to the first rank. The phrase "Chuanzi teaching Jiashan is Chuanzi meeting himself" is the second rank. "Catching the self, catching the hook, being caught by the hook, being caught by the Way" are all expressions of the interplay of apparent opposites.

Although Dogen never explicitly referred to the Five Ranks, he used them often in his writings, engaging them in a way that reflects his depth of understanding and appreciation of their use in practice. In my own commentary on this text, I highlight some of the sections where Dogen used the Five Ranks, showing how he skillfully wove them throughout his teachings.

Through an exhaustive study of the *Mountains and Rivers Sutra* it becomes clear that Dogen's profound insight into the Buddhist teachings is unparalleled in the history of Zen. His understanding of the concept of nonduality and the way that it can inform our everyday actions is of utmost relevance for twenty-first century Zen practitioners and the issues and concerns we are facing, particularly with regard to the environment. Dogen, living as he did in medieval Japan, was certainly not speaking about ecology, global warming, pollution or deforestation. Yet his teachings, when clearly realized, have the potential to guide us in the direction of clear, compassionate activity, whether it's in the realm of environmental activism, social action, or simply in our interactions with each other.

To engage in this kind of study is not an easy matter. The dharma cannot be grasped by linear, sequential thought. It can only be realized in the realm of intimacy, the realm of nonduality. It is there that it needs to function freely so we can respond appropriately to the imperative of our own time and place.

John Daido Loori
Zen Mountain Monastery
Tremper Mountain, New York
Spring 2009

About the Photographs

The photographic images included in this volume are used as capping images for each section of the text. Traditionally, capping phrases or verses are used by Zen practitioners as a way of further expressing the point of a koan. In Japan, the traditional Rinzai Zen koan curriculum includes the use of these capping phrases or *jakugo.* Once monastics have successfully replied to a koan, the Zen master asks them to present a capping verse that expresses their insight in a different way.

The capping images included in this book are not meant to illustrate the text as much as they are intended to illuminate the words of Zen Master Dogen in the same way that capping phrases elucidate the point of a koan. In this sense, they function as an extension of the words and ideas presented in the commentaries and offer another dimension of understanding.

Editor's Note

The poems included at the beginning of each section of commentary are all by John Daido Loori. The excerpts accompanying the photographs are from the sutra itself.

The Way of Mountains and Rivers

Mountains and Rivers Sutra

1. These mountains and rivers of the present are the manifestation of the Way of the ancient sages. Each abides in its own dharma state, exhaustively fulfilling its virtues. Because they exist before the eon of emptiness, they are living in the present. Because they are the self before the appearance of any differences, they are free and unhindered in their actualization. Because the virtues of the mountain are high and broad, the spiritual power to ride the clouds is always entered through the mountains, and the capacity to follow the wind is ultimately liberated from the mountains. Master Dayang Shanggai, addressing the assembly, said, "The blue mountains are constantly walking. The stone woman gives birth to a child in the night." The mountains lack none of their inherent virtues; therefore, they are constantly still and constantly walking. We should dedicate ourselves to a careful study of this virtue of walking. The walking of the mountains is no different than that of humans: do not doubt that the mountains walk simply because they may not appear to walk like humans.These words of the ancient sage Dayang reveal the fundamental nature of walking. Therefore, we should thoroughly investigate his teaching on "constant walking."

2. It is because the blue mountains are walking that they are constant. This walk is swifter than the wind. However, those in the mountains do not sense this, do not know it. To be "in the mountains" is the opening of flowers in the world. Those outside the mountains do not sense this, do not know this. Those without eyes to see the mountains do not sense, do not know, do not see, do not hear this truth. They who doubt that the mountains walk do not yet understand their own walking. It is not that they do not walk, it's just that they do not yet understand, have not yet clarified, walking itself. If we are to understand our own walking we must also understand the walking of the blue mountains. The blue mountains are neither sentient nor insentient; the self is neither sentient nor insentient. Therefore, there should be no doubts about these blue mountains walking.

3. We should realize that the blue mountains must be understood on the basis of many phenomenal realms. We must carefully investigate the walking of the blue mountains, as well as the walking of the self. And this investigation should include walking backward as well as backward walking. We should investigate the fact that since that very time before the appearance of any subtle sign, since the age of the King of Emptiness, walking both forward and backward has never stopped for a moment. If walking had ever rested, sages and wise ones would never have appeared; if walking were limited, these ancient teachings would never have reached the present. Walking forward has never ceased; walking backward has never ceased. Walking forward does not oppose walking backward, nor does walking backward oppose walking forward. This virtue is called "the mountain flowing," it is also called "the flowing mountain."

4. The blue mountains commit themselves to the practice of walking; the East Mountain commits itself to the study of "moving over the water." Hence, this practice is the mountains' own practice. The mountains, unchanged in body and mind, maintaining their own mountain countenance, have always

been traveling about studying themselves. Do not insult the mountains by saying that the blue mountains cannot walk, nor the East Mountain move over the water. It is because of the limitations of the common point of view that we doubt the statement, "the blue mountains walk." It is because of the superficiality of our limited experience that we are surprised by the words, "flowing mountain." Without having fully understood even the words "flowing water," we simply remain mired in ordinary perception. Thus, the accrued merits of the mountain are its name and form, its very lifeblood. There is a mountain walk and a mountain flow, and there is a time when the mountain gives birth to a mountain child. The mountains become sages and wise ones, and it is for this reason that sages and wise ones have thus appeared.

5. Although we may have eyes to see the mountains as the appearance of grass and trees, earth and stones, fences and walls, we should understand that this is not yet their total manifestation; it is not the complete actualization of the mountains. Even when there is a time in which the mountains are perceived as the splendor of the seven treasures, this is still not the real refuge. Even when mountains appear as the realm of the practice of ancient sages, this is not necessarily something to be desired. Even when we see mountains as the actualization of the inconceivable virtue of ancient sages, this is not yet the complete reality. Each of these appearances is the particular objective and subjective result of conditioned views. They are not the actions of the Way of the ancient sages, but narrow and misleading points of view. Turning an object and turning the mind is not condoned by the great sages; explaining the mind and explaining the nature is not confirmed by the sages and wise ones. Seeing the mind and seeing the nature is the business of non-adepts; sticking to words and sticking to phrases is not the expression of liberation. There is expression that is free from such realms; it is "the blue mountains constantly walking," "the East Mountain moving over the water." We should investigate this in detail.

6. "The stone woman gives birth to a child in the night." This means that the time when "the stone woman gives birth to a child" is the "nighttime." Among stones there are male stones, female stones, and stones that are neither male nor female. These stones support the heavens and sustain the earth. There are heavenly stones and earthly stones. Although this is commonly said, it is rarely understood. We should realize the true nature of this "birth." At the time of birth, are both parent and child transformed? We must study and fully understand, not only that birth is actualized in the child becoming the parent, but also that the practice and authentication of the phenomenon of birth occurs when the parent becomes the child. The great Master Yunmen said, "The East Mountain moves over the water." The true significance of this expression is that all mountains are the East Mountain, and each of these East Mountains is also moving over the water. Thus, Mount Sumeru and all of the sacred mountains are actualizing themselves, are all practicing and verifying the truth of "The East Mountain moves over the water." All of this is called "the East Mountain." But is Yunmen himself liberated from the skin, flesh, bones, and marrow of the East Mountain and its life of practice and verification?

7. At the present time in the land of Song there is a group which has grown so large that it cannot be countered by the small number of the genuine students of the Way. They maintain that expressions such as "East Mountain moving over the water" or Nanquan's "sickle" are not rationally comprehensible. They think that any talk which can be grasped by thought is not the Zen talk of the ancient sages and wise ones; indeed, it is precisely incomprehensible talk that is the talk of the sages and wise ones. Consequently, they hold that Huangbo's stick and Linji's roar are beyond logic and unconcerned with thought; they regard these as the great enlightenment that precedes the arising of form before the germination of any subtle sign. They think that the "tangle-cutting words" often used as teaching devices by the great masters of the past are impossible to comprehend. Those who talk in this Way have never met an authentic teacher, and lack the eye of study. What they call "incomprehensible talk" is incomprehensible only to them, not to the sages and practitioners of the Way. Simply because they themselves do not comprehend it is no reason for them not to study the Way that the sages and adepts comprehend. Such people are commonly encountered; they do not know that thought is words; they do not know that words are liberated from thought.

8. We should understand that the teaching of "the East Mountain moving over the water" is the very bones and marrow of the ancient sages. All waters are actualized at the foot of the East Mountain; thus, the mountains ride the clouds and wander through the heavens. The mountains are the peaks of the waters, and in both rising and descending their walk is "over the water." The toes of the mountains walk across the waters, causing the waters to dance; as a result, walking extends freely in the ten directions, and "practice and verification are not nonexistent."

9. Water is neither strong nor weak, neither wet nor dry, neither moving nor still, neither cold nor hot, neither being nor nonbeing, neither delusion nor enlightenment. Solidified, it is harder than diamond: who could break it? When melted, it is softer than milk: who could break it? This being so, how can we doubt the many virtues realized by water? We should reflect on that occasion when the water of the ten directions is seen in the ten directions. This is not a study only of the time when humans or heavenly beings see water: there is also a study of water seeing water. Water practices and validates water; therefore, there is a study of water speaking water. We must bring to realization the path on which the self encounters the self. We must move back and forth along the vital path on which the other studies and fully comprehends the other, and then leap free.

10. In general, there are different ways of seeing mountains and rivers depending on the type of being that sees them. Some beings see what we call water as a jeweled necklace. Yet this does not mean that they see a jeweled necklace as water. How, then, are we to understand what they consider water? Their jeweled necklace is what we see as water. Or, again, they see water as miraculous flowers, though it does not follow that they use flowers as water. Hungry ghosts see water as raging flames or as pus and blood. Dragons and fish see it as palaces and pavilions, or as the seven treasures or the mani gem. Still others see water as woods and walls, or as the nature of immaculate liberation, or as the true human

body, or as the physical form and mental nature. Humans see these as water, and these different ways of seeing bring about the causes and conditions in which water is killed or given life.

11. Therefore, what different types of beings see is different. We should reflect on this. Is it that there are various ways of seeing a single object? Or is it that we have mistaken a variety of images for a single object? We should examine this question in detail, concentrate every effort on understanding it, and then concentrate even more. Given this multitude of perspectives, it follows that training on the way of practice and verification cannot be of one or two kinds, and the realm of ultimate reality must also have a thousand types and ten thousand kinds. If we consider this even further, it seems that although we say there are many types of water, ultimately, there is no original water, no water of various types. However, the various waters which accord with the kinds of beings that see water do not depend on mind, do not depend on body, do not arise from karma, are not self-reliant, and are not reliant upon others. Water, being dependent on water, is liberated.

12. Thus, water is not earth, water, fire, wind, space, or consciousness; it is not blue, yellow, red, white, or black; it is not form, sound, smell, taste, touch, or consciousness. Nevertheless, the water of earth, water, fire, wind, space, is spontaneously being manifested. Because of this, it is difficult to say who is creating this land and palace right now or how they come into being. To say that they rest on the wheel of space and the wheel of wind is not the truth of self nor the truth of other. It is just speculating on the basis of the small view, and is only said out of fear that without such a dwelling place things would not abide. An ancient sage has said, "All things are inherently liberated; they have no abiding place." We should realize that although they are liberated, without any attachments, all things are abiding in their own state. However, when humans look at water they see it only as flowing without respite. This "flow" takes many forms, and the way we see it is just a limited human view. Water flows over the earth; it flows across the sky; it flows up; it flows down. Water flows around bends and into bottomless abysses. It rises to form clouds; it descends to form pools.

13. The Wenzi says, "The Tao of water is to ascend to the sky, forming rain and dew, and to descend to the earth, forming rivers and streams." Such is said even in the secular world. It would be shameful indeed if those who call themselves descendants of adepts and sages had less understanding than secular persons. What this says is that while the path of water is unknown to water, water still actually functions as water; and although the Way of water is not unknown to water, water still actually functions as water. When it "ascends to the sky, it becomes rain and dew." We should realize that water climbs to the very highest heavens, and becomes rain and dew. And this rain and dew is of various kinds in accordance with the various worlds. To say that there are places to which water does not reach is the false doctrine of the lesser teachings. Water extends into flames; it extends into thought, reasoning, and discrimination; it extends into enlightenment and our true nature. "Descending to earth, it becomes rivers and streams." We should realize that when water descends to earth it becomes rivers and streams. And the essence

of rivers and streams becomes sages. Common people think that water is always in rivers, streams, and seas, but this is not so: water makes rivers and seas within water. Therefore, water is in places that are not rivers and seas. It is just that when water descends to earth, it forms as rivers and seas.

14. Furthermore, we should not think that when water has become rivers and seas there is then no world within water: even within a single drop of water incalculable realms are manifested. Consequently, it is not that water exists within these realms, nor that the realms exist within water: the existence of water has nothing whatever to do with the three times or the cosmos. And yet, water is the koan of the actualization of water. Wherever the sages and wise ones are, water is always there; wherever water is, there the sages and wise ones always appear. Therefore, the sages and wise ones have always taken up water as their own body and mind, their own thinking. In this way, the notion that water does not climb up is not found in sacred nor secular writings. The way of water penetrates everywhere, above and below, vertically and horizontally. Still, in the core texts it is said that fire and wind go up, while earth and water go down. But this "up and down" requires further study—the study of the up and down of the Way itself. Where earth and water go is considered "down;" but "down" does not mean some place to which earth and water go. Where fire and wind go is "up." While the cosmic realm has no necessary connection with up and down and the four directions, simply on the basis of the function of the four, five, or six elements we provisionally set up a cosmic realm with directions. It is not that heaven is above and hell below. Hell is the entire universe; heaven is the entire universe.

15. However, when dragons and fish see water as a palace, it's just as when humans see palaces—they do not see them as flowing. And if some bystander was to explain to them that their palace was flowing water they would surely be just as astonished as we are now to hear it said that mountains flow. Still, there would undoubtedly be some dragons and fish who would accept such an explanation of the railings, stairs, and columns of palaces and pavilions. We should consider carefully the reason for this. If our study is not liberated from these confines, we have not freed ourselves from the body and mind of the common person; we have not fully comprehended the land of the ancient sages and adepts; we have not fully comprehended the land of the ordinary person; we have not fully comprehended the palace of the ordinary person. Although humans have a deep understanding of the content of the seas and rivers as water, they do not know what kind of thing dragons, fish, and other beings understand and use as water. We should not foolishly assume that all kinds of beings must use water in the way we understand water. When those who study the Way seek to learn about water, they should not limit themselves to the water of humans; they should go on to study the water of the Way. We should study how we see the water used by the sages and adepts; we should study whether within the house of the sages and adepts there is or is not water.

16. From the timeless beginning to the present, the mountains have always been the dwelling place of the great sages. Wise ones and sages have made the mountains their personal chambers, their own body and mind. And it is through these wise ones and sages that the mountains are actualized. Although many great sages and wise ones have gathered in the mountains, ever since they entered the mountains, no one has encountered a single one of them. There is only the manifestation of the life of the mountain itself; not a single trace of anyone having entered can be found. The appearance of the mountains is completely different when we are in the world gazing at the distant mountains and when we are in the mountains meeting the mountains. Our notions and understanding of nonflowing could not be the same as the dragon's understanding. Humans and gods reside in their own worlds, and other beings may doubt this, or again, they may not. Therefore, without giving way to our surprise and doubt, we should study the words, "mountains flow" with the sages and adepts. Taking one view, there is flowing; from another perspective, there is nonflowing. At one point in time there is flowing; at another, not-flowing. If our study is not like this, it is not the true teaching of the Way.

17. An ancient sage has said, "If you wish to avoid the karma of hell, do not slander the true teaching of the Way." These words should be engraved on skin, flesh, bones, and marrow, engraved on interior and exterior of body and mind, engraved on emptiness and on form; they are already engraved on trees and rocks, engraved on fields and villages. Although it is generally said that mountains belong to the countryside, actually, they belong to those who love them. When the mountains love their master, the wise and virtuous inevitably enter the mountains. When sages and wise ones live in the mountains, the mountains belong to them, trees and rocks flourish and abound, birds and beasts take on a mystical excellence. This is because the sages and wise ones have touched them with their virtue. We should realize that the mountains actually delight in the presence of wise ones and sages. We should understand that the mountains are not within the limits of the human realm or the limits of the heavenly realms. They are not to be viewed with the assumptions of human thought. If only we did not attempt to understand them in terms of flowing in the human realm, who could have any doubts about such things as the mountains flowing or not flowing?

18. Throughout history, we find many examples of emperors and rulers who have gone to the mountains to pay homage to wise ones and seek instruction from great sages. At such times the emperors respected the sages as teachers and honored them without following worldly forms. Imperial authority has no power over the mountain sage, and these emperors understood that the mountains are beyond the mundane world. In ancient times we have the case of the Yellow Emperor who, when he made his visit, went on his knees, prostrated himself, and begged instruction. Again, there were seekers of the truth who left their royal palace and went into the mountains, yet their families felt no resentment toward the mountains nor distrust of those in the mountains who instructed the mountain sages. The years of cultivating the Way are almost always largely spent in the mountains, and it is "in the mountains" that auspicious events inevitably occur. Truly, even a king does not wield authority over the mountains.

19. Since ancient times, wise ones and sages have also lived by the water. When they live by the water they catch fish, or they catch people, or they catch the Way. These are all established water styles. Moreover, going further, there should be catching the self, catching the hook, being caught by the hook, and being caught by the Way. In ancient times, when Chuanzi suddenly left Mount Yao and went to live on the river, he found the sage of Flowering River. Isn't this catching fish? Isn't it catching humans? Catching water? Isn't this catching himself? For someone to meet Chuanzi he must be Chuanzi. Chuanzi's teaching someone is Chuanzi meeting himself. It is not just that there is water in the world; but within the world of water there is a world. This is so not only within water: within clouds there is a world of sentient beings; within wind, within fire, within earth there is a world of sentient beings. Within the phenomenal realm there is a world of sentient beings; within a single blade of grass, within a single staff there is a world of sentient beings. And wherever there is a world of sentient beings, there, inevitably, is the world of the ancient wise ones and sages. We should investigate this truth very carefully.

20. Therefore, water is the palace of the true dragon; it is not flowing away. If we regard it as only flowing, we slander water, for it is the same as imposing nonflowing. Water is nothing but the real form of water just as it is. Water is the water virtue; it is not flowing. In the thorough study of the flowing or the nonflowing of a single [drop of] water, the entirety of the ten thousand realms is instantly realized. Among mountains, there are mountains hidden in jewels; there are mountains hidden in marshes, mountains hidden in the sky; there are mountains hidden in mountains. There is a study of mountains hidden in hiddenness. An ancient wise one has said, "Mountains are mountains and rivers are rivers." This teaching is not saying that mountains are mountains; it says that mountains are mountains. Thus, we should thoroughly study these mountains. When we thoroughly study the mountains, this is the mountain training. Then these mountains and rivers themselves spontaneously become wise ones and sages.

Presented by Eihei Dogen
to the assembly at Kannon Dori Kosho Horin Temple
on the 18th day of the 10th month of the first year of Ninji (1240).

1

These mountains and rivers of the present are the manifestation of the Way of the ancient sages. Each abides in its own dharma state, exhaustively fulfilling its virtues. Because they exist before the eon of emptiness, they are living in the present. Because they are the self before the appearance of any differences, they are free and unhindered in their actualization. Because the virtues of the mountain are high and broad, the spiritual power to ride the clouds is always entered through the mountains, and the capacity to follow the wind is ultimately liberated from the mountains.

Master Dayang Shanggai, addressing the assembly, said, "The blue mountains are constantly walking. The stone woman gives birth to a child in the night." The mountains lack none of their inherent virtues; therefore, they are constantly still and constantly walking. We should dedicate ourselves to a careful study of this virtue of walking. The walking of the mountains is no different than that of humans: do not doubt that the mountains walk simply because they may not appear to walk like humans.

These words of the ancient sage Dayang reveal the fundamental nature of walking. Therefore, we should thoroughly investigate his teaching on "constant walking."

Commentary 1

Where can we put this gigantic body?
When clouds gather on the mountain,
thunder fills the valley.

Throughout the history of civilization, cultures the world over have regarded mountains as sacred places.

Religious pilgrimages and spiritual quests often lead seekers deep into the mountains. The Hindus and Jains travel to Mount Girnar; the Saddhus to Mount Kailash. Spanish monks hike up to the summit of Mount Montserrat; Greek Orthodox priests live on Mount Athos. The Buddha ascended Vulture Peak, Jesus gave his sermon on the Mount, and Moses received the commandments on Mount Sinai. Muhammad was asked to recite the Qur'an in the cave of Hira on Mount Jabal an-Nour. Chinese Buddhists have sought realization on the slopes of Mount Wutai. Dogen built his primary monastery, Eiheiji, deep in the mountains, preferring the unspoiled environment of forested hills, crags and roaring streams to the high society of Kyoto.

In the same way, early on in my teaching career I was called to the mountains. It is there that Dogen's profound teachings contained in the *Mountains and Rivers Sutra* gradually became the guide for practice and training at Zen Mountain Monastery.

But what is the magic and attraction of the mountains? Is there something inherently special in them? If so, what is it?

When we look closely at the mountain we realize we're physically integrated with it. We drink the water that flows out of its springs. We grow our food in what millions of years ago was solid rock. Now it is our garden. We give to the mountain; the mountain gives to us. It becomes part of us. For this reason, it's difficult to say where the mountain ends and we begin.

Once, a friend came up to the Monastery from New York City and I took him on a tour of the grounds. We made our way up a hill bordered by a grove of eastern white pine, past a small pond and around a bend until we reached an open meadow with a magnificent view of the peak of Mount Tremper. My friend stopped dead in his tracks, and staring at the mountaintop he exclaimed:

"Oh, there's the mountain!"

"That's not the mountain," I replied.

My companion stared at me, perplexed. "Then where is it?"

I said, "You're standing on it."

In fact, even to say "standing on" is extra. We *are* the mountain. There is no way that we can separate from it. This being the case, we should ask ourselves, what is the mountain? What are its contours? Where exactly is it?

These mountains and rivers of the present are the manifestation of the Way of the ancient sages. Each abides in its own dharma state, exhaustively fulfilling its virtues. Because they exist before the eon of emptiness, they are living in the present. Because they are the self before the appearance of any differences, they are free and unhindered in their actualization.

In the opening sentence to the *Mountains and Rivers Sutra,* Dogen establishes the fact that mountains and rivers are expressing the teachings of the buddhas and ancient sages, just as a sutra does. In other words, this sutra is not *about* mountains and rivers; it *is* the mountains and rivers. Indeed, if we examine this teaching carefully, we'll see that all phenomena—audible, inaudible, tangible and intangible, conscious and unconscious—are constantly expressing the truth of the universe. A stand of oak saplings, a bed of river rocks, the autumn wind, are all ceaselessly manifesting the Way.

In this paragraph Dogen also says that because the mountains exist before the eon of emptiness—before the appearance of phenomena—they are present here and now. And it is because they live in the present that the self appears and is unhindered in all of its activities.

Master Dayang Shanggai, addressing the assembly, said: "The blue mountains are constantly walking. The stone woman gives birth to a child in the night."

Blue mountains walking and a stone woman giving birth are both inconceivable events. In the context of the dharma, inconceivability points to the inherent emptiness or interdependent origination of all phenomena. Nothing is independent. Nothing has an absolute, own being. And yet, in the relative world, things do indeed exist. There's the child the stone woman gave birth to, there's you, me and the ten thousand things. How do we reconcile this apparent contradiction?

There is an old Zen phrase that says, "When old man Zhang drinks wine, old man Li gets drunk." That is, what happens to you, happens to me. You and I are the same thing, yet I am not you and you are not me. What happens to a snapping turtle in the Catskills, happens to a businessman in Singapore. Yet a turtle is a turtle, a businessman is a businessman.

Whether or not we believe in our identity with all things—in our identity with the mountains and rivers—the fact is that it is the truth of our lives. Belief has nothing to do with it. Understanding has nothing to do with it. We have to realize it. And until we do—until we can clearly see this identity functioning in our lives—we will not truly grasp the effect we have on each other and on this great earth.

These mountains and rivers of the present
are the manifestation of the Way
of the ancient sages.

2

It is because the blue mountains are walking that they are constant. This walk is swifter than the wind. However, those in the mountains do not sense this, do not know it. To be "in the mountains" is the opening of flowers in the world. Those outside the mountains do not sense this, do not know this. Those without eyes to see the mountains do not sense, do not know, do not see, do not hear this truth.

They who doubt that the mountains walk do not yet understand their own walking. It is not that they do not walk, it's just that they do not yet understand, have not yet clarified, walking itself. If we are to understand our own walking we must also understand the walking of the blue mountains. The blue mountains are neither sentient nor insentient; the self is neither sentient nor insentient. Therefore, there should be no doubts about these blue mountains walking.

Commentary 2

Endless blue mountains,
free of even a particle of dust.
Boundless rivers of tumbling torrents,
ceaselessly flowing.

The mountains' walking is an expression of activity in the world. It is because of the mountains' walking—because of their activity—that they are endless. They form a continuum that encompasses past, present and future.

Dogen says, "It is because the blue mountains are walking that they are constant." We could also say that constant walking is the mountains' practice.

Old master Baoche of Mount Mayu was fanning himself. Seeing him, a monastic asked, "Master, the nature of wind is constant and there is no place it does not reach. Why must you still fan yourself?"

Mayu replied, "Although you understand that the nature of the wind is constant, you do not understand the meaning of its reaching everywhere."

The monastic persisted, "What is the meaning of its reaching everywhere?"

Mayu just fanned himself. The monastic bowed with deep respect.

In this koan the monastic is asking, given our original perfection, why do we need to practice? In the early years of Dogen's training, this question was uppermost in his mind. It is what drove him to study exhaustively and to travel to China. It's a question that many of my students ask me: "Since I'm already enlightened, why do I have to do anything?" Because it is through practice that realization and actualization ultimately take place. It is through practice that we must see for ourselves how Mayu's fanning himself is not only the wind reaching everywhere, but the fan, Mayu, the monastic, and us reaching everywhere.

To be "in the mountains" is the opening of flowers in the world. Those outside the mountains do not sense this, do not know this. Those without eyes to see the mountains do not sense, do not know, do not see, do not hear this truth.

"A flower opens and the world arises" is a line from a verse written by the Indian master Prajnatara. This phrase has also been translated as "opening within the world flowers," which means that mountains and the world are one reality. In Zen we use the expression: "In the marketplace, yet not having left the mountain; on the mountain, yet manifesting in the world." Nothing exists outside the mountains. Dogen refers to those "in the mountains" or "outside the mountains," but when he speaks of those "in the mountains," he is not discriminating between those "in the mountains" and the mountains themselves. He is in fact saying that the mountains are identical to those who are "in the mountains." "Those outside the mountains" are also the mountains themselves. The mountain reaches everywhere.

Then there is the line: "Those outside the mountains do not sense this, do not know this." We can understand this in one of two ways. Those outside the mountains are not aware of the mountains walking because they *are* the mountains. They no longer have a reference system with which to see or sense or hear the mountains. Another interpretation is that "those without eyes to see" are suffering from the blindness of ignorance.

In Buddhism there are five kinds of blindness. The first is what we call the blindness of ignorance or separation. The second, the blindness of one who denies the teachings of Buddhism. The third is the blindness of emptiness, where a person first perceives the absolute basis of reality. The fourth is the blindness of attaching to emptiness. The fifth is transcendental blindness, in which there is no distinction between seeing and not seeing.

In the midst of ignorance, we are convinced that we are separate from the mountain and from the rest of the world. We think the world is out there and we are here, contained in this bag of skin. The problem is that when we see ourselves as separate from the rest of the universe, we abdicate our responsibility for it. This is most evident in the way we treat the environment, what we commonly call nature.

In general, we have a very limited understanding of nature. We believe it's made up of phenomena in the physical world but does not include manufactured objects and human interaction. But the fact is that human beings *are* nature—just as much as a tree or a spider web or the Brooklyn Bridge is nature.

How can we discount our own role in creating the earth? We've altered this planet irreversibly. We've produced automobiles, factories, and aerosols. We've refined carbon-based fuels. We've created global warming. All of these are acts of nature—human nature.

Most of the disasters we face today are human-created. Tsunamis and earthquakes kill tens of thousands, but our wars kill millions for profit. Yet we are blind to most of the killing.

When we include the human element in our understanding of nature, we become conscious of the fact that we're responsible for the whole catastrophe. The question then becomes, what will we do about it? When will we do it? What are we waiting for?

To be "in the mountains" is the opening of flowers in the world.
Those outside the mountains
do not sense this, do not know this.

3

We should realize that the blue mountains must be understood on the basis of many phenomenal realms. We must carefully investigate the walking of the blue mountains, as well as the walking of the self. And this investigation should include walking backward as well as backward walking. We should investigate the fact that since that very time before the appearance of any subtle sign, since the age of the King of Emptiness, walking both forward and backward has never stopped for a moment. If walking had ever rested, sages and wise ones would never have appeared; if walking were limited, these ancient teachings would never have reached the present. Walking forward has never ceased; walking backward has never ceased. Walking forward does not oppose walking backward, nor does walking backward oppose walking forward. This virtue is called "the mountain flowing," it is also called "the flowing mountain."

Commentary 3

Though we may speak of it, it cannot be conveyed;
try to picture it, yet it cannot be seen.
When the universe collapses, "it" is indestructible.

Eighteenth century Zen Master Tenkei Denson, who wrote a commentary on the *Mountains and Rivers Sutra,* suggests that Dogen's reference to "walking backward and backward walking" in this passage was actually a mistake; that Dogen really meant to say "walking backward and walking forward." Walking forward is activity, creation. Walking backward is stillness, extinction. Creation and extinction, or activity and stillness, are both characteristics of the blue mountain. When the mountain advances—when you're advancing—activity covers everything. When there is receding, stillness overtakes everything. There is nothing outside of it. But whether the correct translation is "backward walking" or "walking forward," the spirit of what Dogen is pointing to remains the same—whole body and mind stillness and activity.

We must keep in mind that even if we were able to establish without question Dogen's original intention in this or any other passage of the sutra, it would not make his teachings any more understandable or logical. Why? Because his teachings are not meant to be rationalized. They are not intended as an explanation; they're meant to point to the nature of reality.

As Dogen says, we must study the dharma exhaustively—which means we need to practice, realize, and actualize it. There's no other way to plumb its depths.

Once, when I was staying at my teacher Maezumi Roshi's house, I woke up late one night and saw a light in his study. I knocked on the door and heard Roshi's quiet, "Come in." I opened the door to find Roshi sitting at his desk, reading. "Roshi, it's three o'clock in the morning!" I said. "What are you reading?"

"Dogen," he replied.

"Dogen?!" I said, incredulous. He'd been studying Dogen for more than forty years.

He simply looked at me and laughed softly. "Yes, Dogen."

We should investigate the fact that since that very time before the appearance of any subtle sign, since the age of the King of Emptiness, walking forward and backward have never stopped for a moment.

Because all things are intrinsically empty, they lack self nature. To realize this emptiness, the absolute basis of reality, is to realize the first rank of Master Dongshan, "that very time before the appearance of any subtle sign."

The notion of emptiness, or *shunyata,* has become so convoluted in our language that it merits some clarification. First of all, the word "emptiness" is not even an accurate translation.

We use the term in a way that implies that emptiness is the attribute of an object—like roundness. We say that a sphere is round, and simple observation will confirm this perception. Then we apply the same logic to emptiness, describing it as a quality of an object in the phenomenal world. But the emptiness of shunyata is not a thing. It's meant to oppose all views—including the view of emptiness. Shunyata is neither existent nor nonexistent.

When we say that an object is empty, this means that it is empty of independent existence or inherent characteristics. It is interdependent with everything. From a Mahayana Buddhist perspective, emptiness and interdependence are one and the same.

Samadhi, the falling away of body and mind, is usually a practitioner's first experience of emptiness. Samadhi is the state in which we relinquish all our passions and desires. Master Dazhu Huihai said: "Total relinquishment includes all ideas of duality, such as being and nonbeing, love and hate, pure and impure, concentration and distraction. In this, there is no thought of 'Now I see all duality as empty or now I have relinquished all of them.'"

In samadhi, all perspectives disappear. We become completely intimate with the breath, with a koan, with ourselves. Zazen becomes like a bottomless, clear pool.

Whether we are doing zazen or entering the mountains, intimacy is the basis of all of our practice. It is the basis of our lives. If we want to understand the mountain, we have to become the mountain. We have to let the mountain fill our whole body and mind.

A student asked Dongshan, "You always instruct us to follow the way of the birds. What is it to follow the way of the birds?" Dongshan answered, "You don't meet anyone."

What is it to follow the way of the birds, the way of the mountain, the way of the river? What is it to follow our own way? Before we can answer these questions, we must realize that fundamentally, there's not a single thing outside of this gigantic body. In the first rank we realize that our body and mind is the body and mind of the universe.

Master Dongshan's poem on this rank reads:

> In the third watch of the night, before the
> moon appears,
> No wonder when we meet, there is no
> recognition,
> Still cherished in my heart is the beauty of
> the earlier days.

In the evening, before the moon shines there is complete darkness, complete emptiness. It's a state without form, without sound, and without smell. Every last fleck of cloud is wiped from the vast sky. This stage is where the first insight into the absolute basis of reality takes place. It is also the first step in realizing our intimacy with all things.

We should realize that the blue mountains
must be understood on the basis
of many phenomenal realms.

4

The blue mountains commit themselves to the practice of walking; the East Mountain commits itself to the study of "moving over the water." Hence, this practice is the mountains' own practice. The mountains, unchanged in body and mind, maintaining their own mountain countenance, have always been traveling about studying themselves.

Do not insult the mountains by saying that the blue mountains cannot walk, nor the East Mountain move over the water. It is because of the limitations of the common point of view that we doubt the statement, "the blue mountains walk." It is because of the superficiality of our limited experience that we are surprised by the words, "flowing mountain." Without having fully understood even the words "flowing water," we simply remain mired in ordinary perception.

Thus, the accrued merits of the mountain are its name and form, its very lifeblood. There is a mountain walk and a mountain flow, and there is a time when the mountain gives birth to a mountain child. The mountains become sages and wise ones, and it is for this reason that sages and wise ones have thus appeared.

Commentary 4

Let others search for the miraculous
or judge crudeness or fineness.
I only know plum blossoms are white,
azalea flowers are red.

The literature of the world's religious traditions is filled with stories of mystical powers. Buddhism, with its complex metaphysics, is no exception.

One day the Buddha was waiting on a riverbank for a boat to ferry him across the river. An ascetic passed by and proudly showed off his miraculous power, crossing the river back and forth by walking on the surface of the water.

The Buddha, smiling, asked the ascetic, "How long did it take you to attain such power?"

"Thirty years," the ascetic responded, visibly pleased with himself.

"Thirty years!" exclaimed the Buddha. "In only a few minutes, I can cross the river with a boat for a penny."

According to the sutras, walking on water is one of the six spiritual powers, which include supernatural physical prowess, supernatural hearing, knowledge of others' thoughts, recollection of prior lives, supernatural vision, and knowledge of one's own spiritual purification.

The way the Buddha understood and taught mystical power, however, was quite different. He always insisted that this kind of "miraculous power" was, in effect, nothing special. Over a thousand years later, Dogen said the same thing. He insisted that mystical power is not at all supernatural.

Mountains walking on water—when clearly realized—are not supernatural at all. But they are extraordinary. Maintaining a household, raising a child, driving a car, nurturing a relationship—these are the "ordinary" activities most of us spend our daily lives engaged in. Are they any different than mountains walking over the water? How do we define what is ordinary?

The mountains, unchanged in body and mind, maintaining their own mountain countenance, have always been traveling about studying themselves. Do not insult the mountains by saying that the blue mountains cannot walk, nor the East Mountain move over the water.

In Zen we speak of two aspects of training: ascending the mountain and descending the mountain. Ascending the mountain is leaving the world behind and struggling up the mountainside. As we climb, we work with the pain and suffering that we eventually come to recognize as something we have created. Then, when we finally reach the great mystic peak and see the inherent emptiness of all things, we attain realization. But realization is not very useful on top of a mountain, so our training needs to continue down the other side, back into the world. It is there that we begin to see how the absolute basis of reality informs our everyday activity.

Descending the mountain is the second rank of Master Dongshan:

> Sleepy-eyed grandam, encounters herself
> in an old mirror.
> Clearly she sees a face, but it doesn't
> resemble her at all.
> Too bad. With a muddled head she tries
> to recognize her reflection.

At this stage a practitioner emerges out of the absolute basis of reality, where there is no knowing, no consciousness. The moment we realize ourselves, consciousness has already moved. It is on the cusp of awakening that everywhere we go we encounter ourselves.

In the second rank we deal with phenomena in the relative world. In the *Mountains and Rivers Sutra,* this is referred to as mountains walking. It's the continual flux of all things.

You, me, everything we see, think, feel, experience, everything in the whole universe is in a ceaseless state of flux. Nothing is static. This impermanence is the cause of much of our suffering because we cling to things. When we grasp, whatever we're holding on to changes and we change. Yesterday already happened—it doesn't exist. Tomorrow hasn't happened yet—it doesn't exist. This moment and the next moment are different, so what is the use of clinging?

Descending the mountain is by far the most demanding aspect of our practice—much more difficult than realizing the emptiness of phenomena. It's one thing to attain a bit of insight; it's another to actualize it in everything we do—whether it's raising a child, growing a garden, or taking care of this great earth.

In Buddhism we say that all things have a mutual causality, which means all things are intrinsically equal. But to say "all is one" is one-sided. All is not one. War is not the same as peace, organic vegetables are not the same as vegetables grown with the use of pesticides, abuse of the earth's resources is not the same as sustainability. It is clear that all is not one—but neither is it two.

So how do we discern when it is necessary to discriminate, and when we need to be completely intimate with what is in front of us? How do we know when to take action to heal this great earth, and when to defer to its inherent wisdom? Ultimately, there is no rulebook to go by. Each one of us must find our own wisdom. Each one of us must deeply trust ourselves—which means nothing other than giving ourselves permission to really be ourselves.

There is a time when the mountain gives birth to a mountain child.
The mountains become sages and wise ones,
and it is for this reason that sages and wise ones have thus appeared.

5

Although we may have eyes to see the mountains as the appearance of grass and trees, earth and stones, fences and walls, we should understand that this is not yet their total manifestation; it is not the complete actualization of the mountains. Even when there is a time in which the mountains are perceived as the splendor of the seven treasures, this is still not the real refuge. Even when mountains appear as the realm of the practice of ancient sages, this is not necessarily something to be desired. Even when we see mountains as the actualization of the inconceivable virtue of ancient sages, this is not yet the complete reality. Each of these appearances is the particular objective and subjective result of conditioned views. They are not the actions of the Way of the ancient sages, but narrow and misleading points of view.

Turning an object and turning the mind is not condoned by the great sages; explaining the mind and explaining the nature is not confirmed by the sages and wise ones. Seeing the mind and seeing the nature is the business of non-adepts; sticking to words and sticking to phrases is not the expression of liberation. There is expression that is free from such realms; it is "the blue mountains constantly walking," "the East Mountain moving over the water." We should investigate this in detail.

Commentary 5

In arriving, there is no abode;
in departing, there is no destination.
Ultimately, where does it all come down?
Right here, where you have always been.

When Dogen was fourteen years old, he became ordained as a monk at Mount Hiei, the headquarters of the Japanese Tendai School. There he immersed himself in the sutras, the meditation practices of "stopping and seeing," and the esoteric teachings that had been imported from southern India. By the time Dogen was eighteen, he was well versed in the Buddhist canon and could quote from it with ease.

In this section, he presents the four views of the Tendai school and ultimately rejects them as limited ways of perceiving reality.

"The appearance of grass and trees" refers to conditioned origination or the interdependence of all things. It is called the provisional teaching. The "splendor of the Seven Treasures" points to the world of emptiness—the absolute realm—and is referred to as the common teaching. The "realm of the practice of the ancient sages" is the manifestation of conditioned suchness—that is, the nature of phenomena still subject to karma and circumstances. It is called the special teaching. The "actualization of the inconceivable virtue of the ancient sages" represents complete identity with unconditioned suchness. This is the absolute, unchanging, true nature of all things, and it is called the perfect teaching.

But if, as Dogen says, none of these views is yet the ultimate teaching, then what is? In the second paragraph of this section he says:

Turning an object and turning the mind is not condoned by the great sages; explaining the mind and explaining the nature is not confirmed by the sages and wise ones. Seeing the mind and seeing the nature is the business of non-adepts; sticking to words and sticking to phrases is not the expression of liberation. There is expression that is free from such realms; it is "the blue mountains constantly walking," "the East Mountain moving over the water."

Dogen says that "there is expression that is free from such realms; it is 'the blue mountains constantly walking,' 'the East Mountain moving over the water.'" But how is this not turning an object and not turning the mind?

Changsha was once asked by a monastic, "How do you turn the mountains, rivers, and the great earth and return to the self?"

Changsha said, "How do you turn the self and return to the mountains, rivers, and the great earth?"

The monastic's question is turning the mind. Changsha's answer is turning the object.

Subject and object are not two. Mountains, rivers, and the great earth reach everywhere. And whether we try to understand this reality scientifically or spiritually, we cannot deny it. It is the truth of our lives.

If we were to take the water in the spring at Zen Mountain Monastery, for example, and tag every molecule with deuterium—an isotope of hydrogen that can be tracked in water—in a short period of time we would find it in our garden, in the grass, the deer. We would find it in our breath, our urine, in the Esopus river that flows to New York City. Someone coming to the Monastery and eating our food would introduce it into the biological ecosystem of her area. She would keep spreading the deuterium-tagged water until it would eventually reach everywhere. In a short period of time this water would be in the Atlantic Ocean, in Alaska, New Zealand, and Japan.

More importantly, we should understand that this is not only true of water; it is true of all things. It is true of the dharma. There is no place it does not reach.

Explaining the mind and explaining the nature is not confirmed by the sages; seeing the mind and seeing the nature is the business of non-adepts.

Bodhidharma sat in zazen facing the wall. The Second Ancestor, Huike, who had been standing in the snow, cut off his arm and said, "Your disciple's mind is not yet at peace. I beg you, please give it peace."

Bodhidharma said, "Bring the mind to me, and I will set it at rest."

Huike said, "I have searched for the mind, and it is finally unattainable."

Bodhidharma said, "There! I have thoroughly set it at rest for you."

In Buddhism, all of reality is manifested through the three worlds of form, formlessness, and desire. A well-known saying is that the three worlds are nothing but mind, which means that the whole universe is mind. There's not even a speck of dust outside of it. Given this truth, where could we possibly find this mind? How could we see it—let alone explain it? *Who* would explain it? It is only non-adepts who are able to see the mind.

When we define mind in the narrow sense of our intellect, we miss the teachings constantly being proclaimed by the insentient. Most of us go through life oblivious of what is going on around us because our heads are filled with information. It's because we're so educated that we don't have space to receive the teachings of the mountains and rivers. We are constantly talking to ourselves, so how can we hear anything else?

When we are truly able to still our body and mind, we begin to perceive a whole chorus of teachings. Forget all words, forget your ideas, and see for yourself that all things are constantly revealing your nature, the nature of all of beings.

Turning an object and turning the mind is not condoned by the great sages;
explaining the mind and explaining the nature
is not confirmed by the sages and wise ones.

6

"The stone woman gives birth to a child in the night." This means that the time when "the stone woman gives birth to a child" is the "nighttime." Among stones there are male stones, female stones, and stones that are neither male nor female. These stones support the heavens and sustain the earth. There are heavenly stones and earthly stones. Although this is commonly said, it is rarely understood. We should realize the true nature of this "birth." At the time of birth, are both parent and child transformed? We must study and fully understand, not only that birth is actualized in the child becoming the parent, but also that the practice and authentication of the phenomenon of birth occurs when the parent becomes the child.

The great Master Yunmen said, "The East Mountain moves over the water." The true significance of this expression is that all mountains are the East Mountain, and each of these East Mountains is also moving over the water. Thus, Mount Sumeru and all of the sacred mountains are actualizing themselves, are all practicing and verifying the truth of "The East Mountain moves over the water." All of this is called "the East Mountain." But is Yunmen himself liberated from the skin, flesh, bones, and marrow of the East Mountain and its life of practice and verification?

Commentary 6

Parent and child become each other—
they become each other.
Before spring has arrived,
the fragrance of blossoms fills the valley.

All of us enter Zen practice steeped in the world of duality. We're tossed about by events, circumstances, people, and objects. After years of hard work, after taking the backward step of zazen again and again and turning deep into ourselves, we finally reach the place of no differentiation. Eyes open in the middle of the night, we realize the unity of all things. We realize the absolute. This is the "night" that Dogen speaks of—a time of no differentiation. Darkness is complete, emptiness all-encompassing. In this night there is no sight, no sound, no consciousness, no form. "Night" is the absolute basis of reality, the first rank of Dongshan.

In some schools of Buddhism, this is the end of practice, but in Zen, to make a nest in emptiness is just another kind of disease—one that Buddhist practitioners have suffered from for thousands of years. It is tempting to rest in emptiness. In it, there's no conflict, no suffering, no illness or pain. But neither is there activity, consciousness, or even insight. That's why we say that to abide in emptiness is to be a dead person in a coffin with eyes wide open.

Mahayana Buddhism says that once we've reached the peak of the mountain, we must keep going straight ahead, down the other side and back into the marketplace, into the world. Having left the night behind, we enter the world of phenomena in the bright light of day, but this time, our perspective is completely different. We know that form is only one side of emptiness, that emptiness is exactly form.

At the time of birth, are both parent and child transformed? We must study and fully understand, not only that birth is actualized in the child becoming the parent, but that the practice and authentication of the phenomenon of birth occurs when the parent becomes the child.

The child becoming the parent is the relative within the absolute. The parent becoming the child is the absolute within the relative. Here Dogen is essentially encapsulating the first two of the Five Ranks of Master Dongshan, as well as describing the relationship between teacher and student in Zen.

At the beginning of training, master and disciple are very much like parent and child. Students are in completely new territory. They are unsure and in need of fundamental instruction. The job of the teacher is to offer that instruction, much as a parent would to a child. After a while, this relationship changes and the teacher becomes more like a guide, responding appropriately to the students, helping to unstick their sticking points. Then comes the time when the teacher turns into a spiritual friend. Finally, at the time of transmission, teacher and student recognize each other as spiritual equals—the parent becomes the child, the child becomes the parent; the teacher becomes the student, the student becomes the teacher. Dependence on the teacher ends, and the dharma is transmitted to the next generation.

One of the ways in which the identity of teacher and student, of absolute and relative—indeed, of all dualities—is expressed in Zen, is through the practice of liturgy.

Liturgy is a central aspect of our practice. All of Zen's rituals are constantly pointing to the same place: the realization of no separation between the self and the rest of the universe. And because this realization is in truth an act of discovery of what was always there, we say that Zen liturgy "makes visible the invisible."

During the ceremony of dharma transmission, the identity between teacher and student is expressed through a particular kind of liturgy. First, the student circumambulates the teacher sitting on the high seat. Then teacher and student exchange places and the teacher circumambulates the student. The differences between the two are blurred. The child becomes the parent, the parent becomes the child.

Liturgy also has enormous potential for raising our consciousness about our identity with the environment.

In a Zen monastery, *oryoki,* the formal taking of a meal, is a ceremony that points to the fact that, in order to live, we must take life. It's a very detailed liturgy that emphasizes the sacredness of the ordinary activity of eating. But more than just a prescribed form or ritual, oryoki is a state of mind in which we realize our complete interdependence with all things. So is there a way to take this same principle and apply it to our relationship with the environment? Is there such a thing as ecological liturgy?

In 1979, Saint Francis of Assissi was declared the Patron of Ecology, and his "Canticle to the Sun" was adopted by various groups within the environmental movement. Yet when I'm speaking about liturgy, I'm not just referring to prayer. I'm talking about rituals that make us aware—and that make real in our consciousness—the sacredness of this planet. I'm talking about celebrations that express the common experience of a group, religious or otherwise.

We have liturgy for football games, in the court rooms, in family rituals, so why not extend our liturgy to nature, to the insentient?

We should ask ourselves what kind of rituals and celebrations we can do during Christmas, Easter, Passover, Buddha's Birthday, to bring attention to the environment. How can we celebrate Earth Day, not just once a year, but once a month, once a week? Such liturgy would make visible our identity with this great earth.

"The stone woman gives birth to a child in the night."
This means that the time when
"the stone woman gives birth to a child" is the "nighttime."

7

At the present time in the land of Song there is a group which has grown so large that it cannot be countered by the small number of the genuine students of the Way. They maintain that expressions such as "East Mountain moving over the water" or Nanquan's "sickle" are not rationally comprehensible. They think that any talk which can be grasped by thought is not the Zen talk of the ancient sages and wise ones; indeed, it is precisely incomprehensible talk that is the talk of the sages and wise ones. Consequently, they hold that Huangbo's stick and Linji's roar are beyond logic and unconcerned with thought; they regard these as the great enlightenment that precedes the arising of form before the germination of any subtle sign. They think that the "tangle-cutting words" often used as teaching devices by the great masters of the past are impossible to comprehend.

Those who talk in this Way have never met an authentic teacher, and lack the eye of study. What they call "incomprehensible talk" is incomprehensible only to them, not to the sages and practitioners of the Way. Simply because they themselves do not comprehend it is no reason for them not to study the Way that the sages and adepts comprehend. Such people are commonly encountered; they do not know that thought is words; they do not know that words are liberated from thought.

Commentary 7

Clever talk—how can it compare
to the sounds of the river valley,
the form of the mountain?

Dogen's reference to Nanquan's sickle can be traced back to an encounter between Master Nanquan and a monastic as it was recorded in the following koan:

Nanquan was once on the mountain, working. A monastic came by and asked him, "Which is the Way that leads to Nanquan?"

Nanquan raised his sickle and said, "I bought this sickle for thirty cents."

The monastic said, "I am not asking about the sickle you bought for thirty cents. Which is the Way that leads to Nanquan?"

Nanquan said, "It feels good when I use it."

To say that this koan is incomprehensible, and further, that because of its incomprehensibility it must be the true buddhadharma is, from Dogen's point of view, utterly deluded. To think that Linji's roar or Huangbo's stick are beyond logic is equally misguided.

Master Linji, founder of the Linji or Rinzai school of Zen Buddhism, was well known for his shout. A student would ask Linji a question, and he would simply shout in reply. It was a style of teaching that Linji inherited from Huangbo, who in turn received it from Baizhang, one of Mazu's eighty four enlightened successors. Baizhang was also Nanquan's dharma brother.

Linji once said to his assembly:

I see all of you shouting in the East Hall and shouting in the West Hall, don't shout at random. Even if you shout me up to the heavens, break me to pieces and I fall back down again without even a trace of breath left in me, wait for me to revive and I'll tell you it's still not enough. As for all of you here, what are you doing when you just go on with wild random shouting?

In the koans of the various classical collections, Linji, Huangbo, Baizhang, and Mazu are often shown yelling or hitting each other. They overturned chairs, hit monks, and slapped each other. Should we then conclude that this is what Zen is about?

To get caught up in these gestures, to say that they are incomprehensible—and therefore real Zen—is to miss their point. Behind the wild gestures, the seemingly crazy exchanges, there is something these masters were all pointing to.

Many times during my Zen training, I sat in front of my teacher in dokusan on the edge of seeing something, of getting a bit of insight, a bit of clarity. I would ask Roshi a question and then wait breathlessly for his answer. But he did not give me a thing—not a smile, not a twinkle in his eye, not a snort, not a quiver, and yet BAM! something would hit me out of the blue, and I would see it; I would get it.

How was Roshi's silence different from Linji's roar? Was it different?

The same happens now as I work with my students. When they ask me a question, I either point or I stay silent. And they get it. Why? Because they're not receiving anything from me. Their realization emerges from within.

The answers to our questions are not in the shouts, the sticks, the sickles. They're within each one of us. That's why enlightenment cannot be taught. That's why it's called "the wisdom that has no teacher."

Simply because they themselves do not comprehend it is no reason for them not to study the Way that the sages and adepts comprehend. Such people are commonly encountered; they do not know that thought is words; they do not know that words are liberated from thought.

For years, the battle cry of the Linji school was "Painted cakes cannot satisfy hunger." That is, words and ideas cannot reach the truth. Dogen, however, turned this statement upside down and wrote in his *Shobogenzo Gabyo*:

> There is no remedy for satisfying hunger other than a painted rice cake. Without painted hunger you can never become a true person. There is no understanding other than painted satisfaction.

Dogen unequivocally states that outside of painted cakes, there is actually no way to satisfy hunger. That is, words themselves are the true expression of realization. They are realization itself.

Such people are commonly encountered, they do not know that thought is words; they do not know that words are liberated from thought.

Those who think that words are separate from thought do not follow the way of the sages and adepts. They don't know that words are liberated from thought. A vow, for example, is expressed in words yet is not limited to them. To vow means to commit, to practice, to do. When we make a vow, we create karma—action—providing the vow is real. When two people make marriage vows to each other, they create the karma of their particular relationship, which will in turn create karma for others. When we make a vow to do something, that vow is implanted deeply in our psyche. When no one's watching and we transgress, we realize that *we're* watching, and it's painful. Gradually, the gap between what we say and what we do becomes narrower, until the time comes when the gap disappears entirely—until thought, word, and action become one thing.

Such people are commonly encountered; they do not know
that thought is words;
they do not know that words are liberated from thought.

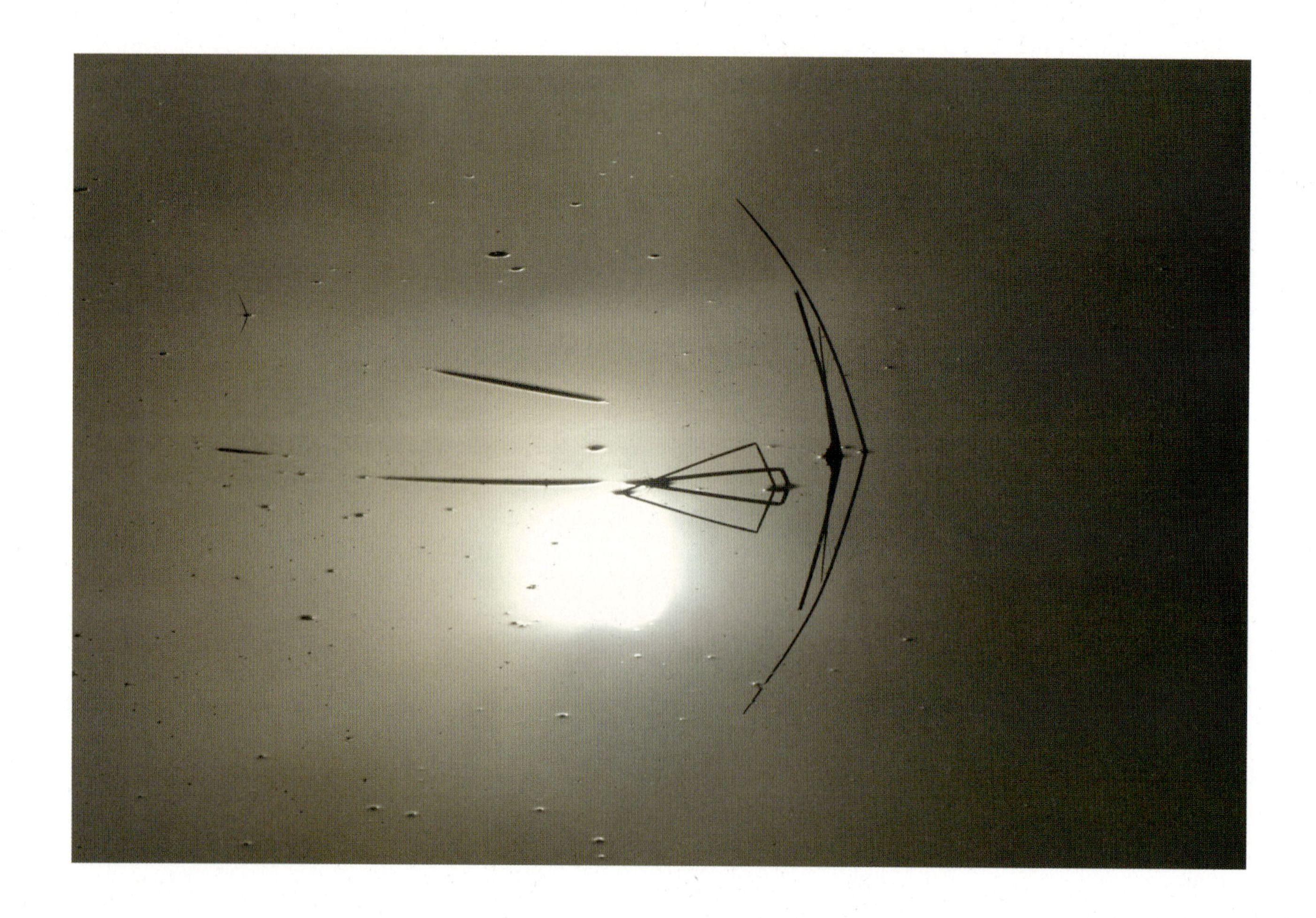

8

We should understand that the teaching of "the East Mountain moving over the water" is the very bones and marrow of the ancient sages. All waters are actualized at the foot of the East Mountain; thus, the mountains ride the clouds and wander through the heavens. The mountains are the peaks of the waters, and in both rising and descending their walk is "over the water." The toes of the mountains walk across the waters, causing the waters to dance; as a result, walking extends freely in the ten directions, and "practice and verification are not nonexistent."

Commentary 8

Blue sky, bright sun—
there is no distinguishing east from west.
Yet acting in accord with the imperative
still requires dispensing medicine
when the sickness appears.

One of the interesting questions that arises when we look at the relationship between form and emptiness is, if these two polarities are completely interpenetrated, why is it that most of us don't live out of this knowledge? Why do we see ourselves as separate from everything and everyone else in the universe?

Discrimination, from the perspective of human evolution, is a necessary trait. In the development of our species, those of us who did not have the wit and skill to deal with a harsh environment were killed by the elements or fell prey to other creatures. Those whose intelligence was developed enough to be able to distinguish an edible from a poisonous plant, or to get out of the way of predators, or make weapons, evolved and proliferated.

The paradox is that discrimination, although necessary, is also at the root of our pain. The very qualities that have helped us to endure as a species: territoriality, dominance, the ability to categorize, are the very qualities that create the suffering all of us deal with throughout our lives.

The Buddha recognized this fact, and out of this recognition the Four Wisdoms or Four Noble Truths came into being. He said that life is suffering; that the cause of suffering is thirst, or desire; that it is possible to put an end to suffering; and that the eightfold path is the means to do it. The path includes: right view, right intention, right speech, right action, right livelihood, right effort, right mindfulness, and right concentration.

In the 2,500 years that followed, different schools of Buddhism developed all sorts of skillful means to help us see that desire arises out of the idea of a distinct and separate self.

But the fact is that inherently, there is nothing wrong with the self. The problem comes when we attach to it, when we think that everything inside the bag of skin is me and everything outside is the rest of the universe.

Xiushan, having returned from a pilgrimage, asked Master Dizang, "I have an unresolved matter, so I'm not willing to go traveling through the mountains and rivers until it's resolved."

Dizang said, "It's not bad that you travel to many mountains and rivers." Xiushan didn't understand Dizang's meaning.

Dizang then asked, "Are the mountains, rivers, and the earth identical or separate from you?"

Xiushan replied, "Separate."

Dizang held up two fingers.

Xiushan hurriedly said, "Identical."

Dizang again held up two fingers.

For a time, Xiushan was lost in thought and then he said, "I don't know whether mountains, rivers and the earth are identical or separate from me."

Dizang asked, "What is it you're calling mountains, rivers, and the earth?"

At this, Xiushan attained realization.

Indeed, what is it that you call yourself? Is this different from mountains, rivers, and the great earth? The self reaches everywhere. It is beginningless and endless. There is nothing outside, nothing we lack, nothing we need to seek. No creature on the face of this earth ever fails to cover the ground upon which it stands.

The toes of the mountains' feet walk across the waters, causing the waters to dance; as a result, walking extends freely in the ten directions, and "practice and verification are not nonexistent."

Zen Master Nanyue went to study with the Sixth Ancestor Huineng, who asked, "Where are you from?"

Nanyue replied, "I came from National Teacher Huian of Songshan."

The Sixth Ancestor asked, "What is it that has come like this?"

Nanyue could not answer. He attended on the master for eight years and worked on the question. One day he met with the Sixth Ancestor, "Now I understand it. When I first came to study with you, you asked me, 'What is it that's come like this?'"

The Sixth Ancestor inquired, "How do you understand it?"

Nanyue answered, "To say it's like something misses it."

The Sixth Ancestor persisted, "Does it depend upon practice and enlightenment?"

Nanyue replied, "It's not that there is no practice and enlightenment. It's just that they cannot be defiled."

Practice and enlightenment cannot be defiled because there is nothing outside of them. Self and other cannot be defiled because there is nothing outside of them. They pervade the whole universe. The same is true of mountains and rivers. Self, other, mountains, and rivers are not "like something." There is no reference system with which to measure them.

At the same time, mountains can be clear-cut, rivers can be drained, and we're the only ones who can stop this from happening. Is this a hopeless endeavor? Sure—so is achieving realization. Yet we'll do it. Clarifying the mind, emptying the mind . . . impossible. But we'll do it.

Practice has nothing to do with hope. Neither does realization. What is required is the kind of tenacity, the kind of vow that comes out of a strong, committed practice.

The toes of the mountains walk across the waters,
causing the waters to dance;
as a result, walking extends freely in the ten directions.

9

Water is neither strong nor weak, neither wet nor dry, neither moving nor still, neither cold nor hot, neither being nor nonbeing, neither delusion nor enlightenment. Solidified, it is harder than diamond: who could break it? When melted, it is softer than milk: who could break it? This being so, how can we doubt the many virtues realized by water? We should reflect on that occasion when the water of the ten directions is seen in the ten directions. This is not a study only of the time when humans or heavenly beings see water: there is also a study of water seeing water. Water practices and validates water; therefore, there is a study of water speaking water. We must bring to realization the path on which the self encounters the self. We must move back and forth along the vital path on which the other studies and fully comprehends the other, and then leap free.

Commentary 9

Pure jewelled eyes, virtuous arms,
formless and selfless, they enter the fray.
The great function works in all ways,
these hands and eyes are the whole thing.

This section is best understood from the perspective of Master Dongshan's third rank, which reflects the development of a certain level of maturity in an individual's practice. Master Dongshan's poem says:

> Within nothingness there is a path
> Leading away from the dusts of the world.
> Even if you observe the taboo
> On the present emperor's name,
> You will surpass that eloquent one of yore
> Who silenced every tongue.

In this stage, a practitioner is able to function for the benefit of others out of the understanding of emptiness and its manifestation in everyday life. This activity is none other than the embodiment of Avalokiteshvara Bodhisattva, the bodhisattva of compassion. Always manifesting according to circumstances, Avalokiteshvara intimately responds to the cries of the world with her ten thousand hands and eyes.

In our culture we have a lot of ideas about compassion, but *karuna*—the Sanskrit term for compassion—is not empathy or sympathy. It's not the same as being nice or doing good. Compassion functions freely, without hesitation, without limitation. It happens without effort, the way we grow our hair, the way we breathe. If others fall, we pick them up. In this kind of activity, there is no sense of doer or doing. Karuna is complete intimacy functioning in the world of differences.

At the center of compassion is the reconciliation of opposites, the merging of differences: self and non-self, sentient and insentient, man and woman, as one reality. Master Dogen calls this merging "I am thus. You are thus." If we examine this statement, it becomes apparent that it is the ground of his metaphysical and religious understanding of great compassion. It is the basis of the non-dual dharma, a state where there is no coming or going, no arising or vanishing.

In another of Master Dongshan's teachings, "Jewel Mirror of Samadhi," he says:

> The mind empty of all activity embraces all
> that appears.
> Like gazing into the jewel mirror, form and
> reflection see each other.
> No coming or going, no arising or vanishing,
> no abiding.
> The ten thousand hands and eyes manifest
> of themselves
> each in accord with circumstances, and yet
> never forget their way.

In this realm there is no holding onto any one place, any one position, any one view.

The great blue heron living in the wetlands comes and goes everyday like clockwork, yet it knows how to step outside of that pattern. It never forgets its own way. The ten thousand hands and eyes of Avalokiteshvara Bodhisattva manifest in accord with circumstances, yet never forget their way.

Water is neither strong nor weak, neither wet nor dry, neither moving nor still, neither cold nor hot, neither being nor nonbeing, neither delusion nor enlightenment. Solidified, it is harder than diamond: who could break it? When melted, it is softer than milk: who could break it?

Dongshan was crossing a river with Yunju when he asked, "Is the water deep or shallow?"

Yunju said, "It's not wet."

Dongshan exclaimed, "That's coarse."

Yunju said, "Tell me, how would you say it?"

Dongshan said, "It's not dry."

Master Dogen says that water is neither strong nor weak, neither wet nor dry, neither moving nor still. Yunju says water is not wet. Dongshan says it's not dry. Then what is it? What is this thing we call water?

This is not a study only of the time when humans or heavenly beings see water: there is also a study of water seeing water. Water practices and validates water; therefore, there is a study of water speaking water.

What does it mean for water to practice and verify water? It means *you* practice and verify yourself, and in so doing, this practice and verification become the practice and verification of all buddhas, past, present and future. Water validating water is the self validating the self. It is the mountains validating the rivers, rivers validating the mountains, mountains validating mountains, and rivers validating rivers.

In *Genjokoan,* Dogen says, "To study the buddha way is to study the self. To study the self is to forget the self. To forget the self is to be enlightened by the ten thousand things." "To forget the self is to be enlightened by the ten thousand things" is the same as seeing the ten thousand things as your own body and mind. Each tip of grass, each dewdrop, each and every thing throughout the whole phenomenal universe contains the totality of the universe.

Pure jewelled eyes, virtuous arms,
formless and selfless, they enter the fray.
The great function works in all ways,
these hands and eyes are the whole thing.

The great function is your life. The ten thousand hands and eyes of great compassion are your hands and eyes. Every time that great heart of compassion comes to life inside your body, you give birth to the hands and eyes of Avalokiteshvara Bodhisattva. That is where she exists—nowhere else.

Solidified, it is harder than diamond: who could break it?
When melted, it is softer than milk:
who could break it?

10

In general, there are different ways of seeing mountains and rivers depending on the type of being that sees them. Some beings see what we call water as a jeweled necklace. Yet this does not mean that they see a jeweled necklace as water. How, then, are we to understand what they consider water? Their jeweled necklace is what we see as water. Or, again, they see water as miraculous flowers, though it does not follow that they use flowers as water. Hungry ghosts see water as raging flames or as pus and blood. Dragons and fish see it as palaces and pavilions, or as the seven treasures or the mani gem. Still others see water as woods and walls, or as the nature of immaculate liberation, or as the true human body, or as the physical form and mental nature. Humans see these as water, and these different ways of seeing bring about the causes and conditions in which water is killed or given life.

Commentary 10

Fundamentally it's not a matter
of interpretation or knowledge.
Summed up, it's not worth a single word.
When the myriad mental activities cease,
truth is miraculously manifested.

Buddhism states that humans are endowed with six senses, not five. In addition to sight, smell, taste, touch, and hearing, mind—or consciousness—is the sixth sense, and its object of perception is thought.

When we look at how we perceive an object, we realize that without consciousness, cognition cannot take place. Sense, object, and consciousness need to function together for us to perceive what we call reality. In order to see this book, for example, your eyes need to be open and functioning, the physical form of the book must be within sight, and your consciousness must be operating. These three components make the reality we call "book." Take any of them away, and the book disappears. If your eyes are closed, you won't see the book. If you take the book and place it somewhere out of sight, it won't exist for you. And if you take consciousness away—as in samadhi, the falling away of body and mind—the book will again remain invisible.

Given this truth, can we say that there is a fixed, unchanging essence of this object we call "book"? Can we say there's an essence of water? Of a self—what Western philosophy would call a soul? If so, where is it? Can you point to it?

Hungry ghosts see water as raging flames or as pus and blood. Dragons and fish see it as palaces and pavilions, or as the seven treasures or the mani gem.

For hungry ghosts—beings with huge, bloated stomachs and throats as thin as needles—water is raging flames because it does not satiate their thirst. For dragons, water is an underground palace where the great serpent, King Nagaraja, guards the teachings of the *Prajnaparamita* literature. Can we deny any of these different kinds of seeing?

Perception depends on a particular reference system. What water means to a polar bear, what it means to a fish, to an eagle, to you, is different. But what happens when we get rid of the reference system altogether? What remains? The whole universe remains. This very body and mind is the body and mind of the universe. But it won't help us to understand this. We need to realize it. We need to make it real in our lives.

Some years ago a rather peculiar man came to the Monastery to sit with us one evening. I was in the dokusan room when the head monitor came to tell me that the new arrival would not sit still. I asked her to send him in. A few minutes later, a middle-aged, ragged fellow walked in and sat down in front of me. "I'm told you're not sitting still in the zendo. If you want to stay, you have to be still."

"I can't help it," the man replied.

"What do you mean you can't help it?"

"They keep sticking me with pitchforks!" he said, visibly upset.

"Who does?!" I asked, surprised.

He pulled up his sleeve and pointed to his bare arm, "They do!" he exclaimed. "The little guys with the pitchforks."

I looked him straight in the eye and said, "Well, tell the little guys they're going to have to sit still in the zendo too! Otherwise, you'll all get thrown out." Then I rang the bell and sent him back to his seat.

Later the monitor told me that our friend had sat like a rock. The next day he returned and I asked to see him again.

"How's it going?" I asked.

"Oh, great, great. Look," he said, smiling as he pulled up his sleeve again. "They're sitting still, just like you said."

Still others see water as woods and walls, or as the nature of immaculate liberation, or as the true human body, or as the physical form and mental nature. Humans see these as water, and these different ways of seeing bring about the causes and conditions in which water is killed or given life.

How humans see water determines whether it is killed or given life.

Fresh water is perhaps the most critical component of any ecosystem, and in many areas of the globe, it is a dwindling resource. Human water consumption rose six-fold in the past century. That's double the rate of the already explosive population growth. We are now using more than fifty per cent of all available fresh water, putting incredible pressure on the environment. Additional demands will further jeopardize all ecosystems. So much water has been taken from rivers that some of them dry up completely before they ever reach the ocean.

For almost thirty years I've been saying that nothing can stop the river in its journey to the great ocean. When I say that I am thinking of a dam. Put up a dam and the river builds up behind it. It goes over or around the dam. Build the dam higher, and the river rises. No matter how high the structure, the river will get past that dam and find its way to the ocean.

It never occurred to me that billions of people could just suck the rivers dry. But that's what's happening all over the world. That's what will keep happening until we do something to stop it.

The environment is all around us. We can't pretend that it's not being affected by our lifestyles. We can't pretend that we don't see our part in its destruction. If we, as spiritual practitioners, are not able to answer the call, then who will?

In general, there are different ways of seeing mountains and rivers
depending on the type of being that sees them.
Some beings see what we call water as a jeweled necklace.

11

Therefore, what different types of beings see is different. We should reflect on this. Is it that there are various ways of seeing a single object? Or is it that we have mistaken a variety of images for a single object? We should examine this question in detail, concentrate every effort on understanding it, and then concentrate even more. Given this multitude of perspectives, it follows that training on the way of practice and verification cannot be of one or two kinds, and the realm of ultimate reality must also have a thousand types and ten thousand kinds.

If we consider this even further, it seems that although we say there are many types of water, ultimately, there is no original water, no water of various types. However, the various waters which accord with the kinds of beings that see water do not depend on mind, do not depend on body, do not arise from karma, are not self-reliant, and are not reliant upon others. Water, being dependent on water, is liberated.

Commentary 11

Appearing without form,
responding in accord with the imperative.
The fragrance of the flower held up
fills the universe existing right here now.

Are there many different kinds of water? Or are there different ways of seeing this single element we call water?

Rain for a farmer who has just sowed the seeds of a new crop is a blessing. For another farmer who is trying to dry his harvested hay, it's a nuisance. For an expert kayaker, the roar of the approaching rapids is ecstatic. For the owner of a house precariously perched at the edge of a riverbank, it's threatening.

How we perceive water completely depends on who we are. It also depends on the time and place in which we find ourselves, as well as our relationship to it.

Years ago, when I was in the Navy, one of my shipmates invited me to visit his home. He was from a Cajun shrimping family outside New Orleans. We traveled in a small motorboat to a wood hut propped up on stilts over the bayou. The brackish smell of the swamp filled the air. Alligators slithered by. Cypress trees emerging from the muck overhung the house.

My friend's relationship to water was very different from anything that I had ever witnessed or experienced myself. The water I was familiar with was always distant, apart from me. He lived in it. Can we say that his perspective was truer than mine? Was his water more real?

Given this multitude of perspectives, it follows that training on the way of practice and verification cannot be of one or two kinds, and the realm of ultimate reality must also have a thousand types and ten thousand kinds.

Given the myriad forms, the ten thousand things that make up this universe, it follows that the way in which we manifest our realization cannot be of only one or two kinds. Each one of us needs to know clearly what we are capable of—how to use our talents and energy for the benefit of others, responding to the imperative of time, place, position, and degree. That appropriate response is at the heart of right action. It is the essence of skillful means, the form that the teachings take in response to circumstances. Skillful means change according to time, place, position, and degree. What was effective twenty years ago may not be effective today. What is skillful in one place may not be appropriate in another.

While I was a research scientist working in a chemical plant, I found out that my company was polluting a local stream. I was in a position of authority at the time, so I used it. I talked to the plant engineer, and although at first he was resistant, when I offered to help him figure out alternative means to get rid of our waste, he agreed to cooperate. We worked together and the pollution stopped. Five years later, when a different ecological crisis involved the same plant, I was no longer working there. All I could do was stand outside the fence with pickets and protest, just like everyone else. My position changed, so my way of dealing with the situation had to change too.

In taking up causes, we also need to be aware of the degree of action required. We have a tendency to fall into extremes. We either wallow in hopelessness, hiding from our problems like an ostrich with its head in the ground, or we run around in a frenzy like a chicken without a head. Either way, we do not accomplish anything worthwhile.

Before Vincent Van Gogh turned seriously to painting, he held a ministry in the Borinage, a coal mining district in Belgium. There he witnessed severe suffering and deprivation. Wanting to help the miners who were part of his congregation, Van Gogh gave away everything he owned: his clothes, food, furniture, even his house. In two weeks he had nothing left. He became one more poor soul shivering in a doorway. Not very skillful at all.

In any given situation, how do we know how much action is necessary? The answer to this question is subtle. Many of us engage a worthy cause with a vengeance. We think we know what's wrong or right, and from that knowing we obtain fuel to propel our anger. But to be able to funnel that energy into effective and skillful action requires that we take all aspects of a situation into consideration. It requires that we be very clear about the many ways in which an object, a person, a place can be seen, so that we can respond to the circumstances appropriately and without self-centeredness.

However, the various waters which accord with the kinds of beings that see water do not depend on mind, do not depend on body, do not arise from karma, are not self-reliant, and are not reliant upon others. Water, being dependent on water, is liberated.

Water is liberated because it is empty of any fixed characteristics. And although there are many ways of seeing water, for the individual being who's seeing it, there is only that one way. Nothing else exists at that moment.

Every single object that we perceive, we have created ourselves. When we look at water, we construct in our minds a particular image that is different from someone else's. Ask a hundred painters to paint the ocean and you'll get a hundred different oceans. But underneath these different images, is there just one ocean?

Michelangelo, when asked to speak about his approach to sculpting, said that he didn't create images; he just released them from the stone. He would patiently chip away until the perfect figure was revealed. Zen practice works in a similar way. It removes all the extra so we can get beyond the multitude of perspectives to the ground of being, so we can realize it and then manifest it in everything we do. This is how we touch everything in this great universe.

Is it that there are various ways of seeing a single object?

Or is it that we have mistaken a variety of images for a single object?

We should examine this question in detail.

12

Thus, water is not earth, water, fire, wind, space, or consciousness; it is not blue, yellow, red, white, or black; it is not form, sound, smell, taste, touch, or consciousness. Nevertheless, the water of earth, water, fire, wind, space, is spontaneously being manifested. Because of this, it is difficult to say who is creating this land and palace right now or how they come into being. To say that they rest on the wheel of space and the wheel of wind is not the truth of self nor the truth of other. It is just speculating on the basis of the small view, and is only said out of fear that without such a dwelling place things would not abide.

An ancient sage has said, "All things are inherently liberated; they have no abiding place." We should realize that although they are liberated, without any attachments, all things are abiding in their own state. However, when humans look at water they see it only as flowing without respite. This "flow" takes many forms, and the way we see it is just a limited human view. Water flows over the earth; it flows across the sky; it flows up; it flows down. Water flows around bends and into bottomless abysses. It rises to form clouds; it descends to form pools.

Commentary 12

The beauty of this garden
is invisible to even the great sages.
The magnitude of this dwelling is so vast,
no teaching can stain it.

The *Heart Sutra,* so named because it is said to be the heart, the distillation of the *Prajna Paramita* literature, is chanted every day in Buddhist monasteries. In it, Avalokiteshvara speaks to Shariputra, one of the Buddha's disciples:

Oh Shariputra, all dharmas are forms of
emptiness;
not born, not destroyed, not stained, not pure,
without loss, without gain.
So in emptiness there is no form; no sensation,
conception, discrimination, awareness;
no eye, ear, nose, tongue, body, mind;
no color, sound, smell, taste, touch, phenomena;
no realm of sight, no realm of consciousness;
no ignorance and no end to ignorance,
no old age and death and no end to old age
and death,
no suffering, no cause of suffering,
no extinguishing, no path, no wisdom, and
no gain.

The *Heart Sutra* says that all dharmas are forms of emptiness, which means that all things are fundamentally without any fixed form. This is true of water, of mountains, of you and me. And yet, all things exist just as they are—thus!

To say that [this land and palace] rest on the wheel of space and the wheel of wind is not the truth of self nor the truth of other. It is just speculating on the basis of the small view, and is only said out of fear that without such a dwelling place things would not abide.

In Buddhist cosmology it is said that beneath the earth there is a set of three disks or "wheels" that support it. Each wheel is composed of the elements of water, wind, and space. Dogen is negating this theory, saying that it's only out of fear that we make such a statement. It is only out of fear that we insist on nailing things down. We want everything in its place. We don't want things to be unfixed or constantly changing. In other words, we don't want them to be the way they are.

From this follows the statement: *An ancient sage has said, "All things are inherently liberated; they have no abiding place."* Things don't rest on the wheel of space or the wheel of wind. They don't rest anywhere. Why? Because they are empty. Yet, water is just water, mountains are just mountains. They have no fixed characteristics, yet they are in their own state.

Although water is inherently empty, it flows up and it flows down. It rises to the sky and rains down on earth. Water becomes dew, ice, and snow. It flows and it's still. It breaks and it melts. Water becomes rivers and streams, lakes and oceans. Water is all of these things, yet it is not any of these things. It is inconceivable.

It is also inconceivable that you and I are the same thing, yet I'm not you and you're not me. It's inconceivable, because our minds are dualistic. We only understand something to be one thing and not another. But the dharma doesn't work that way. Life doesn't work that way. We need to learn to use our minds differently. Or rather, we need to re-learn what we have forgotten after years and years of conditioning. That is the only way that we will be able to see the wonder that surrounds us.

The source of the Hudson River is a small pond, Lake Tear of the Clouds, 4000 feet above sea level on the southwestern slope of Mount Marcy in the Adirondacks. It's a tiny puddle that a kid can jump over, but some 300 miles later it becomes a majestic river discharging 21,400 cubic feet of water per second at the Lower New York Bay in New York City.

The Hudson, originally known as the Tappan Zee—from the name of a local Native American tribe, the "Tappan," and the Dutch "zee" for sea—reflects eastern white pines in the Adirondacks as well as skyscrapers in Manhattan. It is the same river that feeds the Jamaica Bay Wildlife Refuge in Brooklyn. All of its wetlands are dependent on the Hudson.

Because so many of us nowadays are city dwellers, it is easy to romanticize nature at a distance. Sitting in an apartment somewhere in Midtown as we plan a summer adventure to Yellowstone Park, it is important to remember that wild nature is also in our back yard. Despite all of the development going on in the city, Manhattan is still an island. The peregrine falcon and red-tail hawk do not discriminate between a granite cliff or a skyscraper.

Nature exists within and around the city. If we only look, we will see. If we haven't taken off the blinders, we won't see. But that's true in all aspects of our lives. We can either sleepwalk through our days or we can be alive to each moment, each thing. What we choose to do is entirely up to us.

The first step is to notice. The second is to act. Artists like the Hudson River painters, or writers like Annie Dillard, Gary Snyder, and Peter Matthiessen, have devoted their lives to expressing their love for nature. Their work helps us to appreciate the wild as it exhorts us to protect it. Dogen's own poetry often speaks of both the splendor and frailty of nature:

Outside my window, plum blossoms,
just on the verge of unfurling, contain
the spring;
The clear moon is held in the cuplike petals
of the beautiful flower I pick, and twirl.

Even in the dead of winter, as it lies buried beneath three feet of snow, the plum blossom always contains the warmth and life of spring. Each one of us is like that plum blossom. Each one of us is already a buddha, waiting to be awakened. Inconceivable.

We should realize that although they are liberated,
without any attachments,
all things are abiding in their own state.

13

The Wenzi says, "The Tao of water is to ascend to the sky, forming rain and dew, and to descend to the earth, forming rivers and streams." Such is said even in the secular world. It would be shameful indeed if those who call themselves descendants of adepts and sages had less understanding than secular persons. What this says is that while the path of water is unknown to water, water still actually functions as water; and although the Way of water is not unknown to water, water still actually functions as water.

When it "ascends to the sky, it becomes rain and dew." We should realize that water climbs to the very highest heavens, and becomes rain and dew. And this rain and dew is of various kinds in accordance with the various worlds. To say that there are places to which water does not reach is the false doctrine of the lesser teachings. Water extends into flames; it extends into thought, reasoning, and discrimination; it extends into enlightenment and our true nature.

"Descending to earth, it becomes rivers and streams." We should realize that when water descends to earth it becomes rivers and streams. And the essence of rivers and streams becomes sages. Common people think that water is always in rivers, streams, and seas, but this is not so: water makes rivers and seas within water. Therefore, water is in places that are not rivers and seas. It is just that when water descends to earth, it forms as rivers and seas.

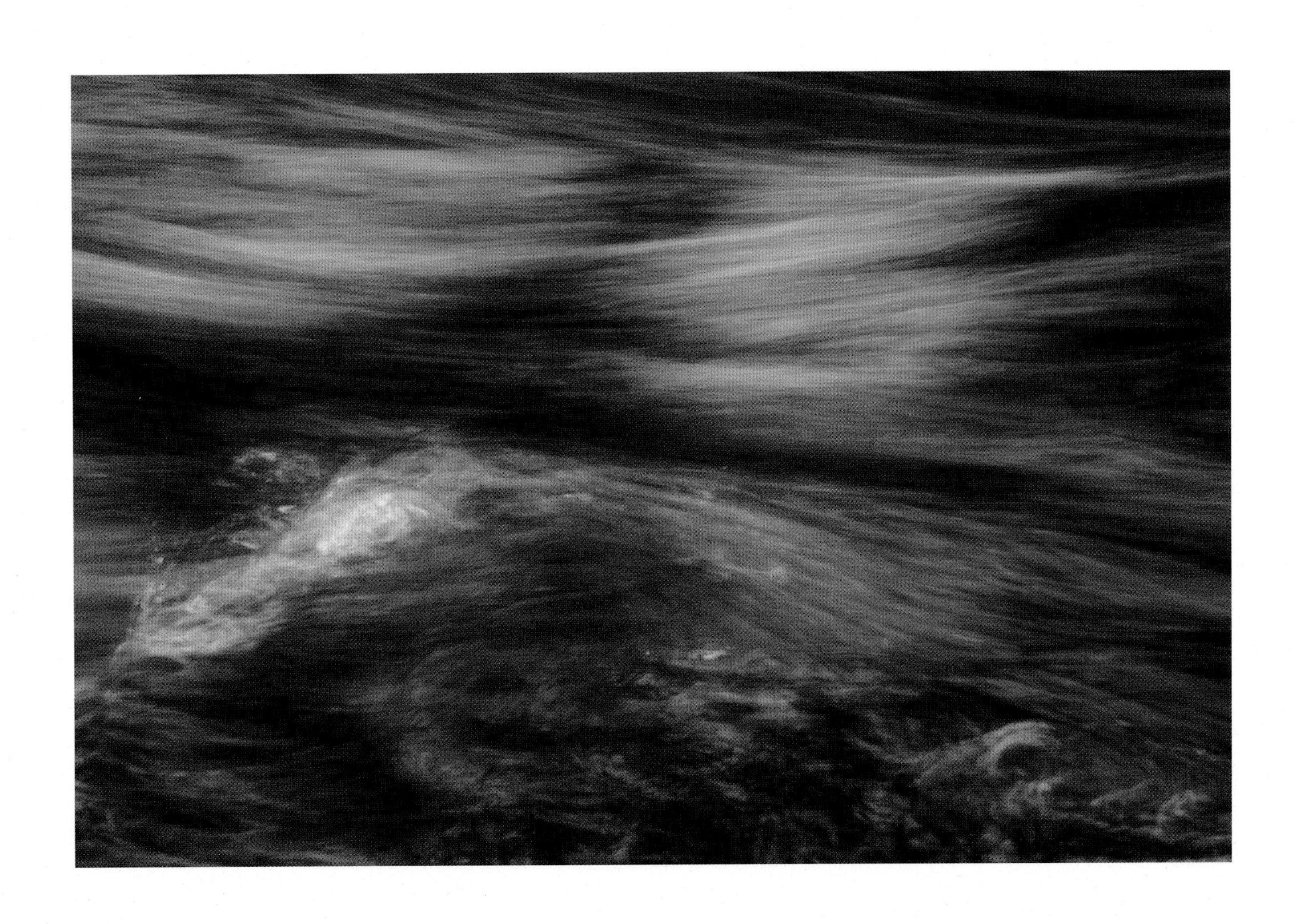

Commentary 13

The dragon howling in a dead tree
clearly sees the Way.
In one, there are many things;
in two, there is no duality.

The *Tao Te Ching* says:

The Way that can be told is not the eternal Way. The name that can be named is not the eternal name.

Emperor Wu of Liang asked the great Master Bodhidharma, "What is the highest meaning of the holy truths?"

The Master said, "Vast emptiness, nothing holy."

The Emperor persisted, "Then who is standing here before me?"

Bodhidharma replied, "I don't know."

In order to know something, we have to be separate from it. When water is intimate with water, it cannot know itself. It's like an eye that cannot see itself, a sword that cannot cut itself. That's why the Way of water is unknown to water. At the same time, the path of water is not unknown to water because water still functions as water.

We should realize that water climbs to the highest heavens, and becomes rain and dew. And this rain and dew is of various kinds in accordance with the various worlds. To say that there are places to which water does not reach is the false doctrine of the lesser teachings. Water extends into flames; it extends into thought, reasoning, and discrimination; it extends into enlightenment and our true nature.

Because water is empty, there are absolutely no restrictions on how or where it appears. Water is in our veins, it's in the wooden pillars that hold up the Monastery building, in its floor tiles, our kitchen pots. There is no place that water doesn't reach.

Sometimes water manifests as a nine-headed dragon, sometimes it manifests as a little baby playing in the garden. Water is flames, it is thought, it is enlightenment.

At one time, Dogen uses water to refer to emptiness, the absolute basis of all things. At another, he holds it up to point to the relative. That is why we cannot get stuck on the words and ideas. We have to go beyond them and directly see the reality itself.

Zen practice is precisely about an immediate experience that is beyond definitions. We can discuss zazen, samadhi, buddha, enlightenment and skillful means. But all of this pales in comparison to the reality of our lives.

Students ask me, "What good is practice? Why am I here? What does Dogen have to do with my life?" They have not yet understood that practice has nothing to do with goals. It has nothing to do with explanations. We don't need a reason to practice. When we've let go of all goals, all reasons and justifications, we're on the Way—and the Way reaches everywhere.

Once, Fayan Wenyi asked Master Dizang for instruction.

Dizang asked, "Where are you going?"

Fayan said, "I am wandering on a pilgrimage."

"What is the purpose of your pilgrimage?"

Fayan said, "I don't know."

"Ah!" Dizang nodded. "Not knowing is most intimate, most intimate indeed!"

At these words, Fayan had realization.

Going to the words and ideas we inevitably end up entangled in a forest of brambles. Old Master Nanquan once said, "The Way is not to be found in knowing or not knowing. Knowing is false consciousness, not knowing is indifference." Then how should we proceed?

At one point in my training at Zen Center of Los Angeles, I was instructing a group of students on liturgy when a particularly eager young man began to pelt me with questions. I launched into a lengthy explanation, when all of a sudden I heard my teacher roar behind me, "That's not liturgy!" In an instant he snatched everything from me.

The truth that Dogen is pointing to cannot be discussed with words. With his unique use of language, he strips away everything that we hold on to. What remains? The wonder of the mystery. We should enter here. Mostly, we shy away from the mystery because it frightens us. We are a civilization that relies on definitions. To enter the mystery is to enter the realm of not knowing and trusting it. It is to trust ourselves.

Common people think that water is always in rivers, streams, and seas, but this is not so: water makes rivers and seas within water. Therefore, water is in places that are not rivers and seas.

The river does not know the ten thousand things that are born from it. It doesn't know the tadpoles, catfish, crayfish, and aquatic snakes. It doesn't know the ducks flocking to a pond and the heron hunting for fish. It doesn't know the bear or the mink or the deer that come to its banks to drink, yet it gives birth to all of them.

We do not need to know the meaning of water in order to know that it quenches thirst. In fact, meaning only gets in the way. When we let go of our opinions, judgments, positions, even river rocks are radiant. When we hold on, even the most precious jewel's brilliance is clouded.

If we are to enter the sacred wonder of the earth we call our home, we have to let go of our preconceived notions. We must hear the song of the river muse, the liturgy of the mountain stream, directly. Only then will their full wonder be revealed to us.

Water extends into flames; it extends into thought,
reasoning, and discrimination;
it extends into enlightenment and our true nature.

14

Furthermore, we should not think that when water has become rivers and seas there is then no world within water: even within a single drop of water incalculable realms are manifested. Consequently, it is not that water exists within these realms, nor that the realms exist within water: the existence of water has nothing whatever to do with the three times or the cosmos. And yet, water is the koan of the actualization of water.

Wherever the sages and wise ones are, water is always there; wherever water is, there the sages and wise ones always appear. Therefore, the sages and wise ones have always taken up water as their own body and mind, their own thinking.

In this way, the notion that water does not climb up is not found in sacred nor secular writings. The way of water penetrates everywhere, above and below, vertically and horizontally. Still, in the core texts it is said that fire and wind go up, while earth and water go down. But this "up and down" requires further study—the study of the up and down of the Way itself. Where earth and water go is considered "down;" but "down" does not mean some place to which earth and water go. Where fire and wind go is "up." While the cosmic realm has no necessary connection with up and down and the four directions, simply on the basis of the function of the four, five, or six elements we provisionally set up a cosmic realm with directions. It is not that heaven is above and hell below. Hell is the entire universe; heaven is the entire universe.

Commentary 14

You and I are the same thing,
yet I am not you and you are not me.
No coming, no going, no arising, no abiding—
as each dharma appears, each dharma is practiced.

The *Avatamsaka* or *Flower Garland Sutra* speaks of the Diamond Net of Indra, a description of the universe in which all things are interconnected, co-arising, sharing a mutual causality. Every node in this net is a diamond with many facets, and each diamond reflects every other diamond. This means that each diamond contains every other diamond.

This net is not a metaphor. It is an accurate description of reality—a description of what all the wise ones and sages have realized, and it will continue to be the realization of all the wise ones and sages in the future. We are totally, completely, intricately interconnected over time and space, with all of time and space.

Shakyamuni Buddha expressed this very same interconnectedness when he said upon his enlightenment: "I and all sentient beings on earth have at once entered the Way." In this statement, "I" is not just the Buddha. All sentient beings on earth arise out of this "I."

When the Buddha was enlightened, all beings were enlightened. In fact, it wasn't just sentient beings—all buddhas, past, present, and future, the mountains and rivers, rocks and trees—attained perfect enlightenment.

Wherever the sages and wise ones are, water is always there; wherever water is, there the sages and wise ones always appear. Therefore, the sages and wise ones have always taken up water as their own body and mind, their own thinking.

The Buddha didn't know he would become the founder of a major world religion. He didn't know about the Diamond Net. He didn't know about koan study or shikantaza. He was just living his life and questioning as he went, trusting himself in the deepest sense.

At every step, he took up each dharma with the whole body and mind. When he was an ascetic, he was the best ascetic on the mountain. When he sat, nobody could outsit him. When he taught, he taught until he dropped in his tracks, combusting for forty-nine years in total dedication. He was not competing with anyone. He wasn't measuring his accomplishments, testing himself against anybody. He didn't retire after ten or twenty years, or after he transmitted to Mahakashyapa. He kept teaching until he couldn't say another word, couldn't take another step. Then, he died—just as completely as he had lived.

That's the way we should live our lives. That's the way we should practice, straightforwardly, with the whole body and mind. When we dance, we should dance with the whole body and mind. When we laugh, we should laugh with the whole body and mind. What else are we willing to settle for?

While the cosmic realm has no necessary connection with up and down and the four directions, simply on the basis of the function of the four, five, or six elements, we provisionally set up a cosmic realm with directions.

Since water reaches everywhere, it cannot go up or down. Dogen is saying that a provisional view based on the four, five and six elements is set up so that we can speak of "up" or "down."

The four elements are earth, water, fire, and wind. The five elements are these four plus consciousness, and the six elements include the element of space.

Although water is empty and does not flow up or down, provisionally we set up a realm with directions so that we can discriminate between water that rises to the heavens and water that flows down into deep abysses.

By now we should realize that Dogen is essentially saying the same thing in different ways. You and I are the same thing, yet I'm not you and you're not me.

In the relative world, water *does* flow up or down. In the relative world, there are massive floods ravaging South Asia and severe droughts wreaking havoc in Australia. Yet, *It is not that heaven is above and hell below. Hell is the entire universe; heaven is the entire universe.* When it floods, it just floods. When it rains, it just rains.

The *Diamond Sutra* says:

An errant monastic does not fall into hell. This very place is the absolute place. There is no hell to fall into. A holy saint does not go to heaven. This very place is the absolute place. There is no heaven to ascend to.

This is the same as saying that because hell is the entire universe, there is no hell to fall into. Because heaven reaches everywhere, there is no heaven to ascend to. Does this mean that we can do whatever we want without consequences? Obviously not.

All of us can appreciate that we must be responsible for our actions—whether it's in the context of our work, our family, or our relationships. What is harder to understand is that the simplest of events also affects the environment. To realize "you and I are the same thing" points to our identity with the whole universe. It also underlines the great responsibility that comes with being human.

Even within a single drop of water incalculable realms are manifested.
Consequently, it is not that water exists within these realms, nor do the realms exist within water.

15

However, when dragons and fish see water as a palace, it's just as when humans see palaces—they do not see them as flowing. And if some bystander was to explain to them that their palace was flowing water they would surely be just as astonished as we are now to hear it said that mountains flow. Still, there would undoubtedly be some dragons and fish who would accept such an explanation of the railings, stairs, and columns of palaces and pavilions. We should consider carefully the reason for this. If our study is not liberated from these confines, we have not freed ourselves from the body and mind of the common person; we have not fully comprehended the land of the ancient sages and adepts; we have not fully comprehended the land of the ordinary person; we have not fully comprehended the palace of the ordinary person.

Although humans have a deep understanding of the content of the seas and rivers as water, they do not know what kind of thing dragons, fish, and other beings understand and use as water. We should not foolishly assume that all kinds of beings must use water in the way we understand water. When those who study the Way seek to learn about water, they should not limit themselves to the water of humans; they should go on to study the water of the Way. We should study how we see the water used by the sages and adepts; we should study whether within the house of the sages and adepts there is or is not water.

Commentary 15

Within darkness there is light;
within light there is darkness.
If you really see it,
you will go blind.

Who is this ordinary person that Dogen speaks of? Is he or she different from the common person? From you and me?

In Zen there is a series of paintings known as "The Ox-herding Pictures" which illustrates the spiritual development of a student from the beginning to the end of formal Zen training. In the first painting, a young practitioner is depicted avidly searching for an ox—the self.

Gradually, the seeker advances on the path, at each step deepening his appreciation of the nature of the self, life, and reality, until he finally arrives at the tenth stage. The tenth ox-herding picture shows a smiling buddha in tattered clothes strolling along without any apparent cares in the world. The sage no longer looks like a sage. He doesn't walk around with a halo; he's just an ordinary being.

A practitioner's spiritual progression is nicely illustrated in the traditional koan collections through the life of the great master Deshan.

It is said that Deshan, a highly-regarded young scholar of the *Diamond Sutra*, heard one day about the special transmission outside the scriptures that was being talked about in southern China, and feeling confident of his own scriptural knowledge and of its importance for Buddhist studies, he set out to disprove the heresy. On the road, Deshan met an old woman who was selling rice cakes.

"What are you carrying in that pack?" asked the old woman.

"My notes on the *Diamond Sutra*," Deshan replied smugly.

The old woman said, "Let me ask you a question. If you can answer it, I'll give you a cup of tea. If you can't, I won't even serve you."

Deshan smiled. "Go ahead, ask."

The old woman said, "In the sutra it says, 'Past mind cannot be grasped, present mind cannot be grasped, future mind cannot be grasped.' If that's true, with which mind will you accept this tea?"

Deshan stared at her dumbfounded.

Despite Deshan's inability to answer, the old woman saw his potential and sent him to study with Master Longtan. When Deshan arrived the two discussed the *Diamond Sutra* all night long. They talked about the nature of mind, of impermanence, of being and non-being. As Deshan was getting ready to leave, he stepped outside and found it was dark. Deshan went back in. "It's dark outside," he said. Longtan lit a candle for him. Deshan took the candle, and as he stepped out into the darkness, the master blew it out. At that moment, Deshan became enlightened. He prostrated himself before Longtan, thanking him profusely. The next day he burned all his notes and said he would never rely on words and ideas again. He then began a pilgrimage to various monasteries.

Deshan traveled across China from east to west, from north to south without saying a word to anyone. Much later, clarifying his understanding further, he became a fearsome teacher. Every time someone asked him a question, he would respond with thirty blows of the stick. All the monks were terrified of him.

But in his eighties, Deshan underwent another transformation.

One evening Deshan went down to the dining hall carrying his bowls.

Xuefeng, who was the Monastery cook, said, "Where are you going, master? The meal bell hasn't rung yet."

Deshan looked at Xuefeng, turned on his heels and meekly walked back to his room.

These incidents show how Deshan moved from being an arrogant know-it-all, to a beginning student of the Way, then a mature practitioner, accomplished teacher, and finally, an ordinary old sage.

Although humans have a deep understanding of the content of the seas and rivers as water, they do not know what kind of thing dragons, fish, and other beings understand and use as water.

In Buddhism, dragons are enlightened beings. They are usually depicted holding a jewel—the mani gem representing the teachings of the dharma—in one of their claws or under their chin. Hakuryusan is a white dragon, guardian of a monastery's buildings and grounds.

As we enter into our second decade of the twenty first century, the need for a legion of bodhisattvas is imperative—a need for enlightened beings from all walks of life to act as the spiritual, environmental warriors of our new millennium. These modern guardians should be individuals of great personal integrity and strong moral and ethical values. They must understand the relationship between religion, politics, the environment and social action, and must be able to fearlessly act out of their understanding. They must be skilled in communication and modern technology and be able to use these skills for the benefit of all beings. Fully appreciating the absolute marvel of the human body and its connection to this great earth, they should know how to nurture, develop, and heal both when necessary. Most importantly, twenty-first century bodhisattvas should know how to creatively express themselves, and to use art as a means of personal and global transformation. In simple terms, these bodhisattvas of today must be able to mediate a conflict, lead a protest, write a poem, love a mountain, nourish a child.

When those who study the Way seek to learn about water,
they should not limit themselves to the water of humans;
they should go on to study the water of of the Way.

16

From the timeless beginning to the present, the mountains have always been the dwelling place of the great sages. Wise ones and sages have made the mountains their personal chambers, their own body and mind. And it is through these wise ones and sages that the mountains are actualized. Although many great sages and wise ones have gathered in the mountains, ever since they entered the mountains, no one has encountered a single one of them. There is only the manifestation of the life of the mountain itself; not a single trace of anyone having entered can be found.

The appearance of the mountains is completely different when we are in the world gazing at the distant mountains and when we are in the mountains meeting the mountains. Our notions and understanding of nonflowing could not be the same as the dragon's understanding. Humans and gods reside in their own worlds, and other beings may doubt this, or again, they may not. Therefore, without giving way to our surprise and doubt, we should study the words "mountains flow" with the sages and adepts. Taking one view, there is flowing; from another perspective, there is nonflowing. At one point in time there is flowing; at another, not-flowing. If our study is not like this, it is not the true teaching of the Way.

Commentary 16

It cannot be described; it cannot be pictured.
The beauty of this garden
is invisible, even to the great sages.
The magnitude of this dwelling is so vast,
no teaching can stain it.

Since time immemorial, sages and wise ones have entered the mountains for periods of fasting, pilgrimage, and retreat, and to build temples and monasteries. Then why does Dogen say that no one has ever met a single one of them?

We should understand that when Dogen speaks of "entering the mountains" he's speaking of the non-dual dharma. There is no separation between the sage and the mountain. When we have made the mountains our own body and mind, our personal chambers, there is no meeting them. Since the mountains and sages are one reality, that the sages have entered the mountains means that there is no one to meet and nothing to be met. There is only the mountain itself.

The appearance of the mountains is completely different when we are in the world gazing at the distant mountains and when we are in the mountains meeting the mountains.

The nature of the mountains is completely different when we separate ourselves from them as observers, and when we are the mountains with the whole body and mind. When we are intimate with something, *it* no longer exists and we no longer exist. There's no way to talk about it, judge it, analyze or categorize it. It fills the whole universe.

Master Dogen said, "To hear sounds with the whole body and mind, to see forms with the whole body and mind, one understands them intimately." To understand intimately does not mean to acquire information. Intimacy is the dwelling place of the great sages. It is realization, a quantum leap of consciousness in which our way of perceiving ourselves and the universe radically changes. And from this new perspective, a different kind of imperative emerges—an imperative for compassionate action.

In Zen we say that enlightenment without morality is not yet enlightenment and morality without enlightenment is not yet morality. On one side, we have the danger of wisdom that lacks compassion. As Gary Snyder once said, "Wisdom without compassion feels no pain." In fact, we could say this is not really wisdom. On the other hand, there is compassion without wisdom, which essentially becomes doing good. Of course, doing good is valuable. But doing good is different from realizing compassion. It is only doing good. When we are doing good, we should carefully examine who or what is being served by our actions. Doing good requires a sense of self: someone who will help someone else. In compassion there is no self, no other, no doer or doing. Ultimately, compassion is dependent on wisdom, wisdom is dependent on compassion. When one arises, both arise. There is no way to separate them. Taking it a step further, we can say that wisdom is compassion, compassion is wisdom. They are two sides of the same reality.

Taking one view, there is flowing; from another perspective, there is nonflowing. At one point in time there is flowing; at another, not-flowing. If our study is not like this, it is not the true teaching of the Way.

A koan in the Miscellaneous Koans of the Mountains and Rivers Order reads:

The bridge flows; the river is still.

What kind of reality is this? We can be sure that it's not something we can explain rationally or understand intellectually. Understanding will only get us so far. Zazen, wisdom and compassion, enlightenment, the Five Ranks—none of these can be understood. They must be realized. They must be made real in everything that we do.

Shakyamuni Buddha never heard of the Five Ranks. Neither did Mahakashyapa nor Ananda. It wasn't until Dongshan and his successor Caoshan decided to create a framework to highlight the interplay between the relative and absolute realms that the Five Ranks came into being. But as is true of any other aspect of the teachings, the Five Ranks are simply a means to realization. They are not an end in themselves.

When it comes down to it, all the skillful means in the world won't help us if we do not realize the mountains, the rivers. They will not help us if we do not realize the nature of the self. Why? Because realizing that the mountains are one's own body and mind is transformative.

"Mountains" are all form—all things, all beings sentient and insentient, and neither sentient nor insentient. To realize all form as one's own body and mind is to dwell in a universe that is unborn and inextinguishable, a universe that has no beginning or end. *You* have no beginning or end. Then how will you care for the mountains and rivers, for your own body and mind, the body and mind of the universe?

The only limits that exist are the ones we have set for ourselves. Take off the blinders and take a step forward. When you've taken that step, acknowledge it, let it go, and take another step. And when you finally arrive at enlightenment, at whole body and mind intimacy, acknowledge it, let it go and take a step forward.

This kind of practice is, always has been, and always will be the ceaseless practice of all the buddhas and ancestors. By practicing in this way, we actualize their very being, their very life. We give life to the Buddha.

The appearance of the mountains is completely different
when we are in the world gazing at the distant mountains
and when we are in the mountains meeting the mountains.

17

An ancient sage has said, "If you wish to avoid the karma of hell, do not slander the true teaching of the Way." These words should be engraved on skin, flesh, bones, and marrow, engraved on interior and exterior of body and mind, engraved on emptiness and on form; they are already engraved on trees and rocks, engraved on fields and villages.

Although it is generally said that mountains belong to the countryside, actually, they belong to those who love them. When the mountains love their master, the wise and virtuous inevitably enter the mountains. When sages and wise ones live in the mountains, the mountains belong to them, trees and rocks flourish and abound, birds and beasts take on a mystical excellence. This is because the sages and wise ones have touched them with their virtue. We should realize that the mountains actually delight in the presence of wise ones and sages.

We should realize that mountains are not within the human realm, nor within the celestial realms. They are not to be viewed with the suppositions of human thought. If only we did not compare them with flowing as understood in the human realm, who could have any doubt about such things as the mountains flowing or not flowing?

Commentary 17

Although adamantine wisdom
is devoid of even a single speck of dust,
how can it compare to sitting alone beneath an empty window,
watching autumn leaves fall, each in its own time.

There are many ways to address the environmental crisis we are facing today. We can rely on legislation to regulate air pollution, water management and waste reduction. We can turn to science to understand the problems we face and apply corrective measures. We can use fear and guilt propagated by the media, and we can look to religion to offer us relevant teachings. However, a much more powerful force than science, legislation, self-interest, and even religion, is love. Regardless of who we are, we all take care of what we love.

Simply spending time in the wilderness can lead us to deeply appreciate its beauty. Environmental organizations, summer camps, and outdoor programs can all foster our love for nature. Nowadays, various kinds of groups offer wilderness immersion workshops that are not just about recreation, but about developing sensitivity to all things wild. Their emphasis is on raising consciousness about our intimate relationship with the environment by emphasizing our identity with it. On a practical level, they encourage the practices of low-impact camping and leaving no traces.

However, environmental research and action, university degrees that focus on ecology and related subjects, green building, and international conferences are only a handful of responses to the dire situation in which we find ourselves. We need a lot more if we're going to save this planet.

Although it is generally said that mountains belong to the countryside, actually, they belong to those who love them.

Long before pollution, overpopulation, and global warming were even recognized as issues, artists in this country understood the importance of raising our consciousness with regards to our relationship with nature.

The environmentalist Jack Turner once said, "Most of us, when we think about it, realize that after our own direct experience of nature, what has contributed most to our love of wild places, animals, plants—and even, perhaps, to our love of wild nature, our sense of citizenship—is the art, literature, myth, and lore of nature."

In the American literary tradition, the writings of poets like Walt Whitman had an enormous impact on people's perception of the wilderness. To this day, it's virtually impossible to read Whitman and not fall in love with nature. There were also the writings of the transcendentalists. In 1841, Emerson wrote an essay called "Thoughts on Art" in which he said: "Painting should become a vehicle through which the universal mind can reach the mind of humankind." In other words, he was encouraging painters to see art as an agent of moral and spiritual transformation. A few years later, the artists of the Hudson River School followed in Emerson's footsteps, creating art the likes of which had never been seen before in the West.

Breaking away from the early influence of the European artists, painters like Thomas Cole, John Vanderlyn, and Asher Durand celebrated the awe and majesty of the Adirondack and Catskill mountains, creating images that went beyond the prevalent view of a human-centered universe. Their sweeping landscapes were reminiscent of the ancient Chinese paintings in which people were portrayed as minuscule dots in a vast, majestic, and unknowable expanse of wilderness.

It's likely that this new way of appreciating nature was linked to the 1895 New York State legislation that designated six million acres of land—the size of the state of Vermont—to remain forever wild. The Adirondack Forest Preserve was the first preserve to be established in the United States, pre-dating the National Forest Preserve by at least ten years.

The Hudson River painters were not alone in their integrative view of nature. Hundreds of years earlier, Zen artists had already recognized the futility of our attempts to dominate the wild. In their *zenga* paintings, their haiku, and the quiet harmony of the tea ceremony, these artist-priests constantly expressed the understanding that we are not only a part of nature, but are actually indivisible from it.

> The plants and flowers
> I raised about my hut
> I now surrender
> To the will
> Of the wind
> —Ryokan

What does it mean to surrender to the will of the mountain? What does it mean to recognize its inherent wisdom? Furthermore, what does it mean to love the mountain, and for the mountain to love its master?

For over thirty years I've watched Tremper Mountain change with the seasons and the winds. I've seen it roar, bolts of lightning flashing over it, trees crashing to the ground, the earth trembling and the river sweeping over rocks and fallen trees. I've also seen it warm and placid, loving, nurturing, and protecting. So what is the true nature of this mountain, of any mountain—indeed, of the whole universe?

This is not an idle question. It is becoming more and more evident that this earth will not tolerate our apathy much longer. But even if we succeed in wiping ourselves off the face of the earth, the planet will eventually renew itself. Given enough time, it will heal. The question is whether or not we will be part of that healing.

Although it is generally said that mountains
belong to the countryside,
actually, they belong to those who love them.

18

Throughout history, we find many examples of emperors and rulers who have gone to the mountains to pay homage to wise ones and seek instruction from great sages. At such times the emperors respected the sages as teachers and honored them without following worldly forms. Imperial authority has no power over the mountain sage, and these emperors understood that the mountains are beyond the mundane world. In ancient times we have the case of the Yellow Emperor who, when he made his visit, went on his knees, prostrated himself, and begged for instruction.

Again, there were seekers of the truth who left their royal palace and went into the mountains, yet their families felt no resentment toward the mountains nor distrust of those in the mountains who instructed the the mountain sages. The years of cultivating the Way are almost always largely spent in the mountains, and it is "in the mountains" that auspicious events inevitably occur. Truly, even a king does not wield authority over the mountains.

Commentary 18

The spiritual potential of the thousand sages
is not easily attained.
Dragon daughters and sons, do not be irresolute—
ten thousand miles of pure wind,
only you can know it.

The monastic Dhrtaka asked Indian Master Upagupta if he could make his home departure.

"Is your intention to renounce the world in body or in mind?" Upagupta asked.

"My request for home departure is not for the sake of body or mind," answered Dhrtaka.

Upagupta pressed him, "Since it's not for body or mind, then who renounces the world?"

Dhrtaka replied, "One who renounces the world has no personal self, no personal possessions, so mentally is neither aroused nor oblivious. This is the eternal Way. The buddhas too are eternal. The mind has no shape or form, and its essence is also thus."

Upagupta said, "You should completely awaken and attain this in your own mind."

Dhrtaka was greatly enlightened.

Leaving home is *shukke-tokudo,* or renunciation of the world. Usually, "home-leaving" is used to describe a student's entrance into monkhood, literally leaving home and family. But home-leaving can also be understood as raising the bodhi mind, and as enlightenment itself. In fact, raising the bodhi mind, leaving home, and enlightenment are one reality.

In our culture, home-leaving is virtually nonexistent. Monastics have jobs, children, homes, and luxury vacations. Lay practitioners hop from retreat center to retreat center looking for a spiritual fix. We're less and less able to give anything up. We want to become enlightened, but we don't want to renounce the world. And what is even worse, we don't realize that everything we attach to helps to build up the layers of conditioning that prevent us from realizing our inherent nature.

One poignant example of the way in which our conditioning obstructs realization is our blatant consummerism.

Some years ago an article appeared in the *New York Times* reporting a burgeoning disease in American society. "It causes mental and emotional deterioration," the article read, "in that it affects the will, commitment, and drive. It causes confusion between what we need and what we want. It weakens vitality, and results in lethargy and an overwhelming lack of spirit. It's called 'Affluenza.'" And apparently, the ones most affected by this condition are children and teenagers. Statistics show that the suicide rate for teens between ten and fourteen years old has more than doubled in the last thirty years. Are we wondering why?

Children that grow up getting everything their heart desires are not very well equipped for life. If anything goes wrong to tip the structure that supports them, they're lost.

One morning, I saw on the business news an interview with an executive from Bell Cosmetics, a company that produces make-up for children ages five through eleven. I watched six-year-olds putting on lipstick and eye shadow, coiffing their hair, applying fingernail and toenail polish while the executive gloated about her multi-billion dollar industry. "Oh, the kids love it!" she said. Then she added, "Statistics show that children ages 6 to 11 have $20 billion to spend." I felt sick. The whole catastrophe was being laid on these unknowing children while the adults delighted in it.

By contrast, the North American Association for Environmental Education launched an initiative called "No Child Left Inside." The program is based on Richard Louv's article, which documents the growing movement to get children to spend more time outdoors.

As you can imagine, these two kinds of children will grow up very differently, with very different consequences.

Whether we like it or not, we must recognize that our behavior is based largely on our conditioning. Bell Cosmetics is just a drop in the bucket. Those children spending their time imitating fashion models will be our business executives twenty-five years from now. They'll be teachers, doctors, and Zen practitioners. What kind of choices will they make? What will be the consequences on the rest of humanity and the planet? What do we, as teachers and parents, want to transmit to them?

Whether we realize it or not, what we do is most indelibly communicated to others, not what we say. Master Dogen called this whole body and mind teaching, and it includes our actions, our silence, our movements, the way we use our minds. In Buddhism we say that karma is created through body, speech, and thought. We can be thinking hate while smiling sweetly, and what we'll communicate is hate and a mixed message. That's why it is so important to be clear about who we are, what our life is. That's why it is so important to see through the layers of conditioning and get to the ground of being.

Every one of us has the spiritual potential of the wise ones and sages, but it's not easy to uncover it. In order to do so, we need to have resolution, commitment, and spirit.

Practice, realization, and actualization are a very personal matter. No one can give us the truth. We have to know it intimately. We have to use it clearly. Instead of greed, we can practice generosity. Instead of attachment we can practice renunciation. Instead of selfishness we can practice selflessness. That's the example we can offer our sons and daughters.

Years of cultivating the Way are almost always largely spent in the mountains,
and it is "in the mountains"
that auspicious events inevitably occur.

19

Since ancient times, wise ones and sages have also lived by the water. When they live by the water they catch fish, or they catch people, or they catch the Way. These are all established water styles. Moreover, going further, there should be catching the self, catching the hook, being caught by the hook, and being caught by the Way.

In ancient times, when Chuanzi suddenly left Mount Yao and went to live on the river, he found the sage of Flowering River. Isn't this catching fish? Isn't it catching humans? Catching water? Isn't this catching himself? For someone to meet Chuanzi he must be Chuanzi. Chuanzi's teaching someone is Chuanzi meeting himself.

It is not just that there is water in the world; but within the world of water there is a world. This is so not only within water: within clouds there is a world of sentient beings; within wind, within fire, within earth there is a world of sentient beings. Within the phenomenal realm there is a world of sentient beings; within a single blade of grass, within a single staff there is a world of sentient beings. And wherever there is a world of sentient beings, there, inevitably, is the world of the ancient wise ones and sages. We should investigate this truth very carefully.

Commentary 

Clouds vanish and the sun appears—
the endless spring is revealed.
This is not a time of yin and yang.
Thus, no communication is possible.
The truth is not like something.

In this section of the sutra Dogen takes up the unique nature of the teacher-student relationship in Zen, probably the most difficult aspect of formal training for Westerners to enter, appreciate, and navigate well.

Mind-to-mind transmission, the central pillar of that relationship, has nothing to do with understanding the teachings or believing a set of dogmas. It is, instead, the recognition of something that was always there.

In ancient times when Chuanzi suddenly left Mount Yao and went to live on the river, he found the sage of Flowering River. Isn't this catching fish? Isn't it catching humans? Catching water? Isn't this catching himself? For someone to meet Chuanzi he must be Chuanzi. Chuanzi's teaching someone is Chuanzi meeting himself.

Here Dogen is refering to a koan in which Jiashan, already the abbot of a temple, realizing that there was something missing in his understanding of the dharma, went to see Chuanzi, the sage of the Flowering River:

. . . Seeing Jiashan approaching, Chuanzi said to him, "Of which monastery are you the abbot?" Jiashan said, "I am not abbot of a monastery, or I wouldn't look like this." Chuanzi said, "What do you mean by 'not like this'?" Jiashan said, "It's not like something right in front of you." Chuanzi said, "Where did you study?" Jiashan said, "No place that ears or eyes can reach." Chuanzi said, "The phrase you understand can still tether the donkey for a myriad kalpas." Then he said, "I hang a line one thousand feet deep, but the heart is three inches off the hook. Why don't you say something?" Jiashan was about to open his mouth. Chuanzi knocked him into the water with the boat pole. Jiashan surfaced and climbed onto the boat. Chuanzi said, "Say something. Say something!" Jiashan was about to open his mouth when Chuanzi hit him again. Jiashan was suddenly awakened and bowed three times.

Chuanzi said, "You're welcome to the fishing pole. However, the meaning of 'it ripples no quiet water' is naturally profound." Jiashan said, "Why do you give away the fishing pole?" Chuanzi said, "It is to see whether a fish of golden scales is or is not. If you have realized it, speak quickly; words are wondrous and unspeakable." While Chuanzi was speaking, Jiashan covered his ears and began to walk away. Chuanzi said, "Quite so, quite so." Then the master instructed Jiashan: "From now on, erase all traces, but do not hide your body. I was at Yaoshan's for thirty years and clarified just this. Now you have this. Do not live in a city or village. Just be in a deep mountain or on a farm and guide one or half a person. Succeed in my teaching and don't let it be cut off." Jiashan accepted Chuanzi's entrustment and bid him farewell. He went ashore and started to walk away, looking back again and again. Chuanzi called out, "Reverend, reverend!" Jiashan turned around. Chuanzi held up the oar and said, "There is something more." Upon uttering these words, he jumped out of the boat and disappeared into the mist and waves.

This excerpt is part of a difficult yet very powerful koan. In the *Eihei Koroku,* another collection of Dogen's teachings, he comments on this case: "Although when Jiashan was at the temple he was excellent in discussion, he expounded the teachings to humans and celestials, he was perfect in speech and no one could defeat him in argument, it still wasn't complete." Jiashan had everything that any fine abbot would have, and still, his practice was incomplete. After seeing Chuanzi, he realized himself, so Chuanzi offered him the fishing pole. Handing over the fishing pole is similar to handing over the *kutz,* the *shippei,* or the staff, everyday objects that later became part of the formal transmission ceremony. With this gesture, Chuanzi was affirming Jiashan. And this, Dogen says, is Chuanzi meeting Jiashan, Jiashan meeting Chuanzi. It is also Chuanzi meeting himself, and Jiashan meeting himself.

In order for the teacher to meet the student, the teacher must meet herself, the student must meet himself. The teacher must meet the student, the student must meet the teacher. This is the "crossing of two blade points," as Master Dongshan says in his verse to the fourth rank:

> When two blades cross points,
> There's no need to withdraw.
> The master swordsman
> Is like the lotus blooming in the fire.
> Such a person has a heaven-soaring spirit.

The lotus is a symbol of enlightenment, fire is the symbol of delusion. In the fourth rank of Master Dongshan—the rank of mutual integration—enlightenment and delusion, teacher and student, the lotus and the fire, merge. The lotus doesn't exist outside of the fire. In fact, it's because the fire burns that the lotus can bloom. A good practitioner is like a lotus blooming in a fire. When he gives life, he gives life through and through, always acting according to conditions.

So in this koan, what is it that Chianzi, and then Jiashan realized? What was it that was transmitted between them? Chuanzi almost drowned Jiashan and Jiashan got it. Old Huangbo punched out his disciple Linji and Linji got it. What kind of crazy practice is this? It has nothing to do with the gestures or the forms. It's not about walking three feet off the ground or ecstatic, cosmic experiences. Then, you tell me, what is it really about? What is it that Jiashan realized, and why did Chuanzi say, "Reverend, there is something more"?

Since ancient times, wise ones and sages have also lived by the water.
When they live by the water
they catch fish, or they catch people, or they catch the Way.

20

Therefore, water is the palace of the true dragon; it is not flowing away. If we regard it as only flowing, we slander water, for it is the same as imposing nonflowing. Water is nothing but the real form of water just as it is. Water is the water virtue; it is not flowing. In the thorough study of the flowing or the nonflowing of a single [drop of] water, the entirety of the ten thousand realms is instantly realized.

Among mountains, there are mountains hidden in jewels; there are mountains hidden in marshes, mountains hidden in the sky; there are mountains hidden in mountains. There is a study of mountains hidden in hiddenness. An ancient wise one has said, "Mountains are mountains and rivers are rivers." This teaching is not saying that mountains are mountains; it says that mountains are mountains. Thus, we should thoroughly study these mountains. When we thoroughly study the mountains, this is the mountain training. Then these mountains and rivers themselves spontaneously become wise ones and sages.

Commentary 

Aimlessly, the spring breeze, of itself,
knows how to enter
the scars of the burning.

Arriving at the end of this profound sutra we can now understand some of Dogen's teachings within the framework of the Five Ranks of Master Dongshan.

The first of the Five Ranks describes the absolute basis of reality. The second rank is emergence out of the realization of the absolute. The third rank is the manifestation of that realization in the relative world. It's a synthesis of form and emptiness, where compassion begins to manifest effortlessly. The fourth rank is mutual integration—it is Avalokiteshvara bodhisattva ceaselessly responding to the cries of the world.

The first two ranks show both sides of all phenomena—absolute on one side, relative on the other—while recognizing the relationship between the two. In the third and fourth ranks, the two realms are independent. Absolute is absolute, and relative is relative. A devil is a devil, a buddha is a buddha.

Finally, Dogen brings it home to the fifth rank. Here, complete unity is attained, so that unity and disparity, form and emptiness, absolute and relative, all disappear. Everything is seen at once, and no trace of enlightenment or non-enlightenment remains.

Dongshan's verse to the fifth rank reads:

Who dares to equal the one
Who falls into neither being nor non-being!
All of us want to leave
The current of ordinary life,
But this one, after all, comes back
To sit among the coals and ashes.

Dogen makes a similar point in his *Genjokoan:* "No trace of enlightenment remains and this traceless enlightenment continues endlessly." We call this endless activity "filling a well with snow," the seemingly inane occupation of the ancient sages. No one can tell whether they're wise or crazy, ordinary or holy. One of them gathers a few others and they all climb the mountain to get to the snow-capped peaks. They fill their buckets with snow and they carry them down and throw the snow into the well, trying to fill it. Of course, this is impossible. Yet they do it, trip after trip, day after day. It is like the Four Bodhisattva Vows that all Zen practitioners make each night:

Sentient beings are numberless; I vow to save them.
Desires are inexhaustible; I vow to put an end to them.
The dharmas are boundless; I vow to master them.
The Buddha Way is unattainable; I vow to attain it.

It is impossible to save numberless beings, yet we vow to do it. Impossible to exhaust inexhaustible desires. Impossible to master infinite dharmas. Impossible to attain the unattainable. Impossible, yet we'll do it.

An ancient wise one has said, "Mountains are mountains and rivers are rivers." This teaching is not saying that mountains are mountains; it says that mountains are mountains.

An ancient master once said, "Thirty years ago, before studying Zen, I saw mountains as mountains and rivers as rivers. When I had more intimate knowledge, I came to see mountains not as mountains and rivers not as rivers. But now that I have attained the substance, I again see mountains just as mountains, and rivers just as rivers."

The zazen of a beginner is innocent. It's free, open, and receptive. But after a while, it becomes rote. It's one thing to really practice this incredible Way with the whole body and mind, and quite another to simply look like a Zen practitioner. Much of our practice involves maintaining this freshness, this receptivity.

This teaching is not saying that mountains are mountains; it says that mountains are mountains.

This is the mountain of the nature of all dharmas, the ten thousand things, the whole phenomenal universe. It pervades all time and space, from the beginningless beginning to the endless end. It is the body and mind of the ten thousand things—and, it's just a mountain.

Thus, we should thoroughly study these mountains. When we thoroughly study the mountains, this is the mountain training. Then these mountains and rivers themselves spontaneously become wise ones and sages.

When Dogen says "thoroughly study the mountains," he means to take these mountains and rivers as the koan of our lives. Whether we look at these mountains and rivers with the eyes of a biologist, a sage, a deer, as the mountain, or as the river, the fact is that they are constantly proclaiming the dharma. The river sings the eighty-four thousand verses. Do we hear them? The mountain reveals the form of the true dharma. Can we see it?

When we go deep into ourselves, when we engage Zen practice fully, that practice becomes the practice of all buddhas past, present, and future. It is the verification and actualization of the enlightenment of Shakyamuni Buddha and all of the subsequent buddhas. It is also the practice and verification of these mountains and rivers, and of your life and my life, the life of wise ones, sages, and ordinary beings.

Deeply realizing ourselves and the true nature of these mountains and rivers is perhaps the most important and profound thing each of us will ever do with our lives. We should not take it lightly. Our lives and the life of this planet depend on it.

When we thoroughly study the mountains, this is the mountain training.
Then these mountains and rivers themselves
spontaneously become wise ones and sages.

Index of Photographs

Page 9 Way of Water
Page 11 Frozen Mountain
Page 14 Colorado Mountains
Page 15 Manifesting the Way
Page 17 Endless Mountains
Page 20 In the Mountains
Page 21 Hudson River Palisades
Page 23 Horseshoe
Page 26 Delaware River Bank I
Page 27 Delaware River Bank II
Page 29 Raquette Lake Shoreline
Page 32 Mountain Jewels
Page 33 Desert Form
Page 35 Water of the Way
Page 38 Water Sculpture I
Page 39 Water Sculpture II
Page 41 Birth
Page 44 Flourishing Rock
Page 45 Rock and Water Reflections
Page 47 Morning Mist
Page 50 Water Grasses
Page 51 Raindrops
Page 53 Mountain Clouds
Page 56 Shadows
Page 57 Water Dance
Page 59 Harder than Diamond
Page 62 Still Point
Page 63 Ice Mushroom
Page 65 Water of the Ten Directions
Page 68 Water Extends into Flames
Page 69 Water Validating Itself
Page 71 Water Seeing Water
Page 74 Water Fulfilling Its Virtues
Page 75 Dancing Waters
Page 77 Delaware Falls
Page 80 No Abiding Place
Page 81 Abiding in Its Own State
Page 83 Water Extending Itself
Page 86 Descending to Earth
Page 87 Water Verifying Water
Page 89 A World within Water
Page 92 Incalculable Realms
Page 93 Liberated Water
Page 95 Ice Head Form
Page 98 Winged Waves
Page 99 Water Flowing
Page 101 Blue Heaped on Blue
Page 104 Distant Mountain
Page 105 Hovering Clouds
Page 107 Mani Gem
Page 110 Mountain Dawn
Page 111 Reflections
Page 113 Hidden in Hiddenness
Page 116 Ice Mountains
Page 117 Delaware River Bank III
Page 119 Green Flow
Page 122 Blue Flow
Page 123 Wave Edges
Page 125 Softer than Milk
Page 128 Evening Stillness
Page 129 Receding Flow

Glossary of Names and Buddhist Terms

Avalokiteshvara bodhisattva of compassion, sometimes described as having one thousand hands and eyes. Her name means "One who hears the sounds (of the sufferings of the world)."

backward step turning one's attention inward; studying the self through Zen meditation and practice.

Bodhidharma an Indian monk known for taking Buddhism from India to China, where he settled at Shaolin monastery and practiced meditation for nine years facing the wall.

bodhisattva one who compassionately postpones final enlightenment for the sake of others.

buddha nature according to Mahayana Buddhism, the enlightened and immutable nature of all beings.

dharma universal truth or law; the Buddha's teachings; all phenomena that make up reality.

Deshan (780 – 865, C.E.), a successor of Longtan, he taught at Mount De; he was known for his fierceness as a teacher and his "thirty blows" style of answering questions.

Diamond Sutra key text of the *Prajñaparamita* literature; it states that phenomenal appearances are illusory projections of the mind, empty of self.

Dongshan (807 – 869, C.E.), a successor of Yunyan. Together with Caoshan he is regarded as the founder of the Caodong or Soto School, one of the five schools of Chinese Zen.

enlightenment realization; the direct experience of one's true nature.

emptiness central principle of Buddhism that recognizes that all things are without self nature; often equated with the absolute in Mahayana, since it is formless and non-dual.

dokusan private face-to-face interview with the teacher during which students present and clarify their understanding of the teachings.

falling away of body and mind *samadhi;* state in which the mind is absorbed in intense concentration, free from distractions and goals; single-pointedness of mind.

Flower Garland or Avatamsaka Sutra Mahayana sutra that constitutes the basis of the Huayan school; it states that buddha, mind, and universe are identical to one another.

Genjokoan *The Way of Everyday Life;* the first fascicle and the heart of Master Dogen's *Shobogenzo.*

ground of being buddha nature; the true nature of the self.

Hakuin Ekaku (1689 – 1769, C.E.), is known as the "revitalizer" of the Rinzai School of Zen because of his systematization of koan training and his emphasis on the importance of strong zazen practice.

insentient i.e., mountains, rocks, and tiles; beings which still have the ability to express the dharma.

karma universal law of cause and effect; it equates the actions of body, speech, and thought as potential sources of karmic consequences.

koan an apparently paradoxical statement used in Zen to induce in students an intense level of doubt, allowing them to cut through conditioned reality and see directly into their true nature.

Linji (? – 866, C.E.), a successor of Huangbo and founder of the Linji school of Zen, one of the two schools of Buddhism still active in Japan.

Mahayana "Great vehicle;" the northern school of Buddhism that expresses and aims at the intrinsic connection between an individual's realization and the simultaneous enlightenment of all beings.

Manjushri bodhisattva of wisdom, depicted with the sword that "kills and gives life."

mind-to-mind transmission confirmation of the merging of the minds of teacher and disciple; also the recognition of the buddha mind and entrustment of the teaching.

Nanquan (748 – 834, C.E.), a successor of Mazu and teacher of Zhaozhou; he is well known for his koan "Nanquan Kills a Cat."

non-dual dharma essential principle of existence; nondiscrimination or the lack of dualistic opposition.

practice also ceaseless practice; according to Master Dogen, a continuous process of actualizing enlightenment.

Samantabhadra protector of those who teach the dharma; also known as the Great Conduct Bodhisattva.

Shakyamuni Buddha Siddhartha Gautama, the historical Buddha and founder of Buddhism; he was a prince of the Shakya clan, living in northern India in the sixth century B.C.E.

shikantaza "just sitting;" form of zazen in which meditators practice pure awareness.

Shobogenzo *Treasury of the True Dharma Eye;* Master Dogen's masterpiece consisting of ninety-two fascicles or writings on various aspects of the dharma.

skillful means (S. *upaya*) forms that the teachings take, reflecting their appropriateness to the circumstances in which they appear.

suchness tathata; the absolute, immutable, true state of phenomena.

sutra narrative text consisting mainly of the discourses and teachings of the Buddha.

Tendai Buddhism school based on the *Lotus Sutra* and the teaching of the three truths: emptiness, temporal limitation of existence, and suchness.

True Dharma Eye complete English translation of *Master Dogen's Three Hundred Koan Shobogenzo* by Kazuaki Tanahashi and John Daido Loori with Loori's commentary, capping verses and footnotes.

Way, the practice of realization taught by Shakyamuni Buddha; the nature of reality.

Yunmen (864 – 949, C.E.), a successor of Xuefeng and founder of the Yunmen School, one of the five schools of Chinese Zen.

About John Daido Loori

John Daido Loori is the abbot and resident teacher at Zen Mountain Monastery in Mt. Tremper, New York, as well as the founder and director of the Mountains and Rivers Order. Loori is a successor to Hakuyu Taizan Maezumi Roshi and is trained in both the vigorous school of koan Zen and the subtle teachings of Dogen's Zen. Devoted to maintaining the authenticity of these traditions, Daido Loori has, over the past three decades, developed a distinctive style of teaching and an eight-point training program that is geared equally to lay and monastic practitioners. With zazen and a strong teacher-student training relationship at its core, the program takes place within a uniquely American context and addresses not only religious activities and concerns but virtually every aspect of daily life.

Daido Loori is also an award-winning photographer and videographer. He has exhibited widely, both in the United States and abroad, and his photographs have been published in *Aperture* and *Time Life*. He is the author of twenty-four books, some of which have been translated into several languages. His most recent titles are *The True Dharma Eye*, *Sitting With Koans*, and *The Zen of Creativity*.

About Zen Mountain Monastery and the Mountains and Rivers Order

The Mountains and Rivers Order is a Western Zen Buddhist lineage that includes various organizations dedicated to the support of authentic and engaged spiritual practice.

Zen Mountain Monastery, the Order's main house, is a monastic center providing traditional yet distinctly American Zen training. Recognized as one of the leading Zen Buddhist monasteries in the West, it offers year-round Zen training under the guidance of its founder and abbot, John Daido Loori Roshi. MRO teachers also include Konrad Ryushin Marchaj Sensei, vice-abbot of Zen Mountain Monastery, and Geoffrey Shugen Arnold Sensei, resident teacher and vice-abbot of Zen Center of New York City.

The Society of Mountains and Rivers is an umbrella organization for affiliates associated with the Monastery. This network of sitting groups in the United States and abroad provides facilities and visiting teachers, encouraging consistent zazen and projects that bring spiritual practice into every area of life.

The Zen Environmental Studies Institute is a not-for-profit religious corporation that provides education and training in Zen and its relationship to the environment.

Dharma Communications is the Monastery's educational outreach arm, and it provides the general public with audiovisual materials, electronic media and publications on Buddhism, arts, sciences, ethics, social action, business and ecology.

Recognizing the wealth of Buddhist teachings and the importance of preserving them for future generations, the National Buddhist Archive was created to collect and organize materials of enduring value related to the history and activities of the Mountains and Rivers Order, and to make these materials available to the general public for the purpose of Buddhist study.

For more information on the Mountains and Rivers Order, visit our website at www.mro.org.